TM 9-785

WAR DEPARTMENT TECHNICAL MANUAL

HIGH SPEED TRACTOR M4 TECHNICAL MANUAL

DECLASSIFIED
Auth: NND 785097
By NNHP - N~~~~

Dissemination of restricted matter. The information contained in restricted documents and the essential characteristics of restricted materiel may be given to any person known to be in the service of the United States and to persons of undoubted loyalty and discretion who are cooperating in Government work, but will not be communicated to the public or to the press except by authorized military public relations agencies. (See also paragraph 18b, AR 380-5, 28 September 1942.)

UNCLASSIFIED

REGRADED _____ BY AUTHORITY

WAR DEPARTMENT　　　　　　　　　　　*FEBRUARY 1943*

©2011 Periscope Film LLC
All Rights Reserved
ISBN #978-1-937684-96-9

This book has been digitally watermarked by
Periscope Film LLC to prevent
illegal duplication.

TM 9-785

RESTRICTED

WAR DEPARTMENT

TECHNICAL MANUAL

18-TON HIGH SPEED TRACTOR M4

1 NOVEMBER 1943

Dissemination of restricted matter.—The information contained in restricted documents, and the essential characteristics of restricted materiel, may be given to any person known to be in the service of the United States, and to persons of undoubted loyalty and discretion who are cooperating in Government work, but will not be communicated to the public or to the press except by authorized military public relations agencies. (See also paragraph 18b, AR 380-5, 28 September 1942.)

*TM 9-785

TECHNICAL MANUAL
TM 9-785

RESTRICTED

WAR DEPARTMENT
Washington, 1 November 1943

18-TON HIGH SPEED TRACTOR M4

Dissemination of restricted matter.—The information contained in restricted documents, and the essential characteristics of restricted materiel, may be given to any person known to be in the service of the United States, and to persons of undoubted loyalty and discretion who are cooperating in Government work, but will not be communicated to the public or to the press except by authorized military public relations agencies. (See also paragraph 18b, AR 380-5, 28 September 1942.)

CONTENTS

PART ONE—OPERATING INSTRUCTIONS

			Paragraphs	Pages
SECTION	I	Introduction	1-2	3-6
	II	Description and tabulated data	3-4	7-10
	III	Driving controls and operation	5-17	11-22
	IV	Operation of auxiliary equipment	18-21	23-29
	V	Operation under unusual conditions	22-24	30-35
	VI	First echelon preventive maintenance	25-29	36-46
	VII	Lubrication	30-31	47-51
	VIII	Tools and equipment stowage on tractor	32-39	52-58

PART TWO—VEHICLE MAINTENANCE INSTRUCTIONS

SECTION	IX	Maintenance allocation	40-41	59-66
	X	Second echelon preventive maintenance	42	67-82
	XI	Organization tools and equipment	43	83
	XII	Trouble shooting	44-56	84-107
	XIII	Engine	57-63	108-126
	XIV	Ignition system	64-67	127-130

*This technical manual supersedes TM 9-785, 1 February 1943, TB 785-1, 12 August 1943, TB 785-2, 13 August 1943, and TB 785-3, 18 August 1943.

18-TON HIGH SPEED TRACTOR M4

			Paragraphs	Pages
Section	XV	Fuel system	68-74	131-138
	XVI	Intake and exhaust systems	75-78	139-144
	XVII	Engine cooling system	79-84	145-159
	XVIII	Engine lubricating oil filters and oil cooler	85-87	160-164
	XIX	Electrical system	88-97	165-192
	XX	Instruments and gages	98-111	193-204
	XXI	Master clutch	112-114	205-207
	XXII	Torque converter and propeller shaft	115-118	208-219
	XXIII	Transmission, differential, and final drives	119-121	220-223
	XXIV	Steering brakes	122-124	224-234
	XXV	Air trailer brake controls	125-129	235-246
	XXVI	Electric trailer brake controls	130-131	247-254
	XXVII	Tracks and suspensions	132-137	255-274
	XXVIII	Winch and power take-off	138-143	275-284
	XXIX	Ammunition and cargo boxes	144-145	285-287
	XXX	Pintle	146-147	288-289
	XXXI	Gun ring, fire extinguisher, and hull drain	148-150	290-291

PART THREE—STORAGE AND SHIPMENT

Section	XXXII	Shipment and temporary storage	151-153	292-296
	References			297-298
	Index			299

PART ONE—OPERATING INSTRUCTIONS

Section I

INTRODUCTION

	Paragraph
Scope	1
Arrangement	2

1. SCOPE.

a. This technical manual is published for the information and guidance of the using arm personnel charged with the operation and maintenance of this materiel.

b. In all cases where the nature of the repair, modifications, or adjustment is beyond the scope of facilities of the unit, the responsible ordnance service should be informed so that trained personnel with suitable tools and equipment may be provided, or proper instructions issued.

2. ARRANGEMENT.

a. In addition to a description of the 18-ton High Speed Tractor M4, this manual contains technical information required for the identification, use, and care of the materiel. The manual is divided into two parts. Part One, section I through section VIII, gives vehicle operating instructions. Part Two, section IX through section XXXII, gives vehicle maintenance instructions for using arm personnel charged with the responsibility of doing maintenance work within their jurisdiction.

18-TON HIGH SPEED TRACTOR M4

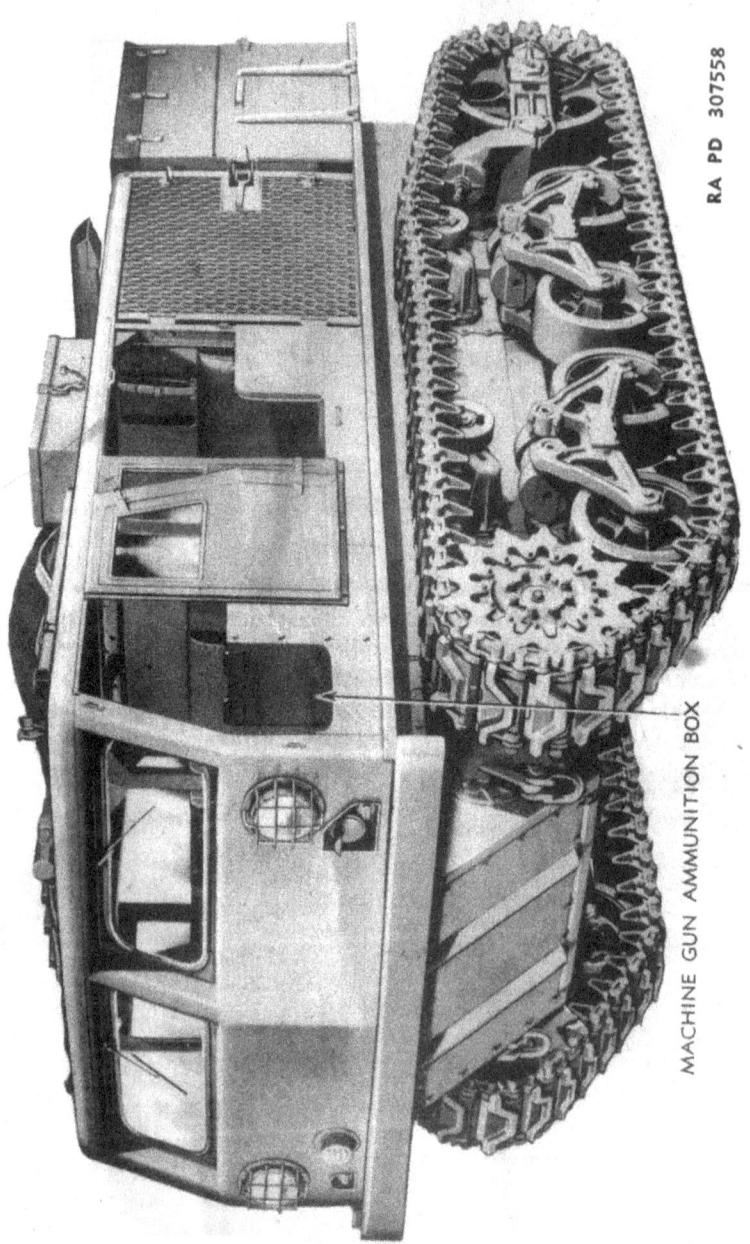

Figure 1 — Left Front View of Vehicle

Figure 2 — Right Rear View of Vehicle

TM 9-785

18-TON HIGH SPEED TRACTOR M4

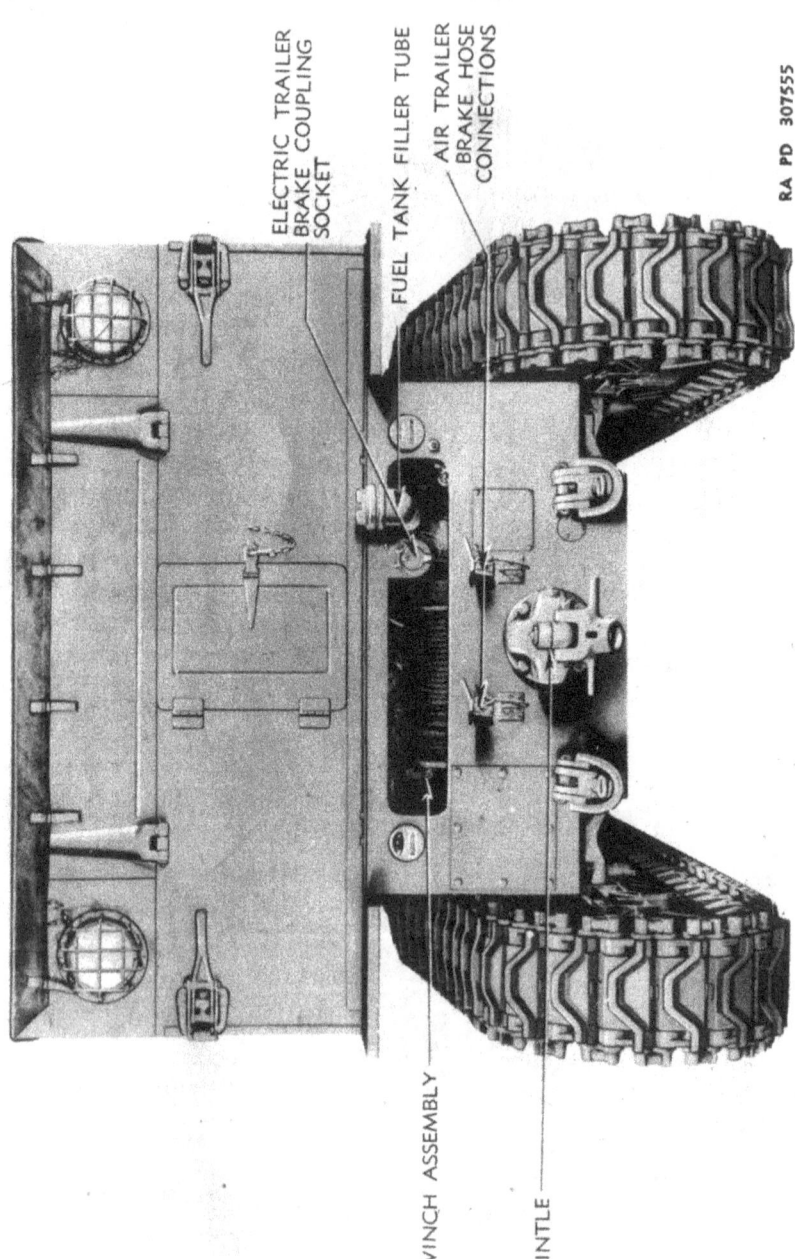

Figure 3 — Rear View of Vehicle

TM 9-785

Section II

DESCRIPTION AND TABULATED DATA

	Paragraph
Description	3
Tabulated data	4

3. DESCRIPTION.

a. **General.** This tractor is designed primarily as a prime mover for 3-inch and 90-mm antiaircraft gun mounts, 155-mm gun carriages, and 8-inch and 240-mm howitzer carriages. Except for the cargo compartment for carrying ammunition and gun accessories, all tractors regardless of the load they tow are identical. To avoid confusion it is at all times essential to designate the standard basic vehicle as 18-ton High Speed Tractor M4 and to follow this designation with the caliber of the weapon in parentheses. Thus proper identification can be made of vehicles with cargo compartments designed to carry different types of ammunition and gun accessories. The tractor is of the crawler or track-laying type and may be used for either highway or cross-country travel, or where mountains, swamps, sand, or small unbridged trenches may be encountered. It has three forward speeds and one reverse speed and because of its low center of gravity, the tractor can easily climb slopes which may be as steep as 30 degrees, depending on the kind of footing available and the load being pulled. An engine brake gear provides for holding tractor to a low speed when descending steep hills.

b. **Engine.** Power is supplied by a 6-cylinder, water-cooled, 4-cycle, valve-in-head gasoline engine with a minimum rated horsepower of 210 at 2100 revolutions per minute.

c. **Steering.** Steering is accomplished by means of a controlled differential. The action is controlled by means of brakes and hand levers.

d. **Seats.** The cab is divided into two compartments. The front compartment seats the driver and two other men, while the rear compartment accommodates eight men on two facing rows of seats. The seat cushions are padded canvas zipper bags; back cushions are leather-covered.

e. **Equipment.** Equipment on the tractor includes batteries and electrical ignition and lighting system, electric starter and generator, speedometer and engine tachometer, fire extinguisher, electric and air trailer brake controls, and winch. A gun ring in top of cab provides for mounting of a caliber .30 or caliber .50 machine gun. An ammunition or cargo box at the rear of the tractor provides for carrying of dunnage and ammunition. Tools are carried in boxes and brackets on the tractor top.

18-TON HIGH SPEED TRACTOR M4

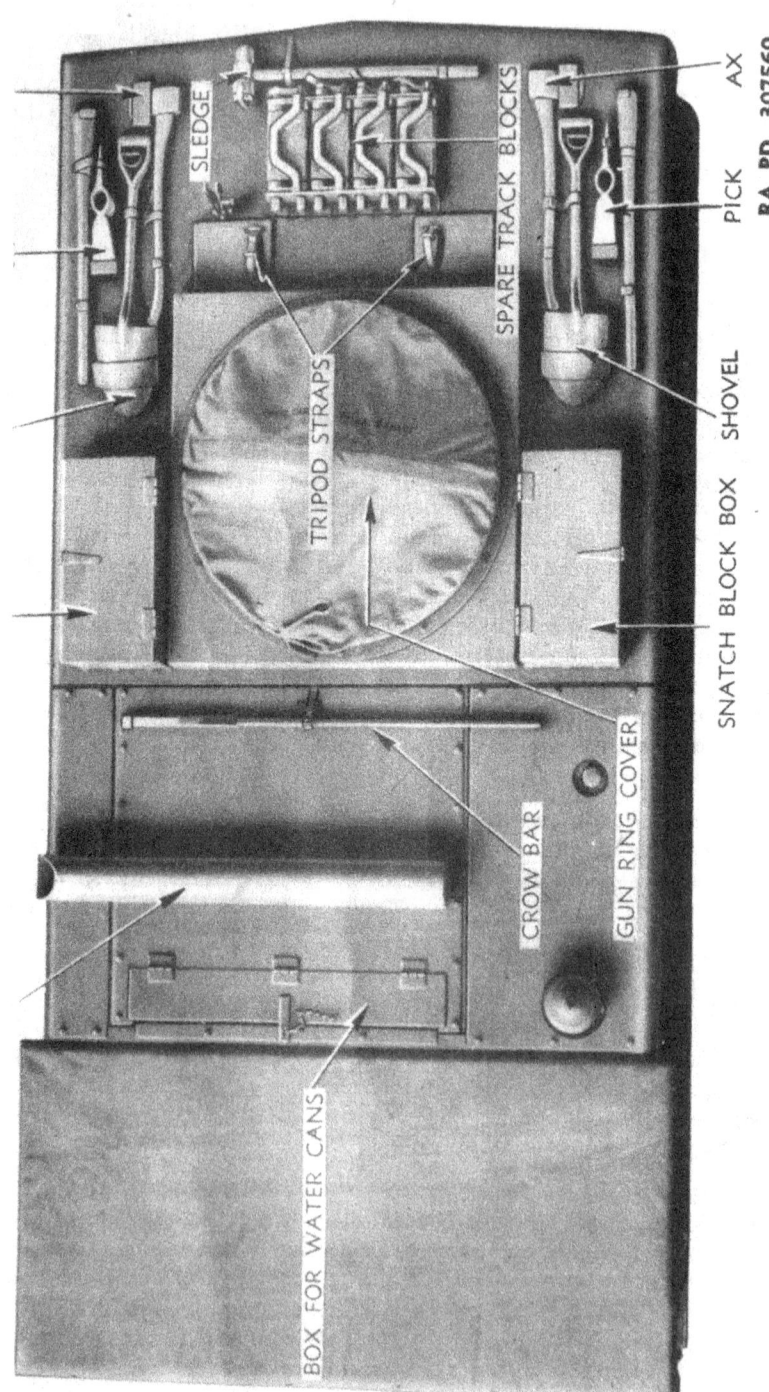

TM 9-785
3-4

DESCRIPTION AND TABULATED DATA

f. Tractor and Engine Numbers. Engine number will be found on a plate on the right side of the engine block (viewing engine from flywheel end). Tractor serial number is stamped on plate in cab and also on rear of tractor above and to the right of right-hand taillight.

4. TABULATED DATA.

a. General.

Maximum drawbar horsepower at 2100 rpm (full throttle):
  Creeper gear ....146.6 at 4 mph
  Low gear ....140.8 at 6 mph
  High gear ....105.6 at 14 mph
Maximum drawbar pull:
  Creeper gear ....38,700 lb at stall
  Low gear ....13,000 lb at 4 mph
  High gear ....4,160 lb at 9 mph
Weight (without crew or ammunition box estimated).......28,000 lb
Weight (with crew and ammunition box estimated).........31,400 lb

b. Dimensions.

Over-all length ....16 ft 11 in.
Over-all width ....8 ft 1 in.
Over-all height ....8 ft 3 in.
Tread width (center to center of tread)...................6 ft 8 in.
Ground clearance ....1 ft 8 in.
Height of pintle ....2 ft 5 in.

c. Performance.

Speeds and drawbar pull at 2100 rpm (full throttle):
  Creeper gear ....7 mph, 6,410 lb
  Low range ....9 mph, 5,000 lb
  High range ....35 mph, 200 lb
  Reverse ....8.3 max. mph
Maximum trench crossing (estimated).........................5 ft
Maximum fording depth (at minimum speed)..............3 ft 5 in.
Maximum grade-ascending ability (estimated)..............30 deg
Maximum grade-descending ability (estimated)............30 deg
Allowable list (side slope—estimated) ....30 deg
Maximum vertical obstacle ....2 ft 5 in.
Turning radius ....18 ft 6 in.

d. Engine.

Make and model ....Waukesha—145 GZ
Number of cylinders 6
Firing order (counting from end of engine closest to
  rear of tractor) 1-5-3-6-2-4
Bore and stroke ....$5\frac{3}{8}$ x 6 in.

9

18-TON HIGH SPEED TRACTOR M4

Piston displacement .... 817 cu in.
Rated speed (rpm at full throttle) 2,100
Maximum torque .... 585 ft-lb at 1,500 rpm

   e. **Steering.**

Method .... Controlled differential
          (by means of brakes and hand brake levers)

   f. **Tracks.**

Ground contact area .... 4,140 sq in.
Length of track on ground .... 10 ft 4 in.
Width of shoes .... 16 9/16 in.
Number of shoes per track 65
Ground pressure (lb psi) .... 7.6 lb with 90-mm gun
Ground pressure (lb psi) .... 8.75 lb with 155-mm gun
Type of bearings in bogie wheels and idlers .... Tapered roller

   g. **Winch.**

Length of drum .... 18 in.
Diameter of drum .... 6½ in.
Length of cable .... 300 ft
Diameter of cable .... ¾ in.
Maximum cable speed (bare drum) ... 55 fpm at 1,400 rpm of engine
                              (do not exceed)
Maximum pull .... 30,000 lb

   h. **Power Take-off.**

Maximum revolutions at rated engine
  speed .... 1,000 rpm at 1,400 rpm of engine
                              (do not exceed)

   i. **Capacities.**

Fuel tank .... 125 gal
Engine crankcase only .... 18 qt
Engine crankcase with filter change .... 6¼ gal
Engine cooling system .... 18 gal
Transmission case .... 7 gal
Final drive cases (each) .... 10 qt
Winch gear housing .... 3 qt
Torque converter system .... 8½ gal
Air cleaner .... 5 qt
Torque converter pump housing .... 1 qt
Trailing idler .... 3 pt
Bogie .... 1½ pt
Track support roller .... 1 pt
Fan drive shaft housing .... ⅓ qt
Fan drive gear housing .... ⅞ qt

Section III

DRIVING CONTROLS AND OPERATION

	Paragraph
Inspection of new tractor	5
Prepare new tractor for use	6
Operation of new tractor	7
Engine starting instructions	8
Stopping engine	9
Use of master clutch and gearshift lever	10
Steering levers and brakes	11
Towing of vehicle	12
Use of trailer brakes	13
Parking vehicle	14
Lighting system	15
Hour meter	16
Instruments	17

5. **INSPECTION OF NEW TRACTOR.**

a. Make a complete inspection for any shortage or damage which may have occurred while in transit or storage. Check tools and equipment in, or with, the tractor against the list given in section VIII to make sure nothing is missing.

6. **PREPARE NEW TRACTOR FOR USE.**

a. The battery ground cable is disconnected after tractor is loaded at the factory and must be connected to negative terminal of battery before tractor can be started. Refer to paragraph 89 j for instructions for connecting cable to battery. Add water to battery if level is not ⅜ inch above separators (par. 89 d). Tractors consigned for overseas shipment are shipped with dry charged battery and electrolyte for battery is shipped with each tractor in a separate container. Before starting tractor, the electrolyte must be put into battery according to instructions given on tag fastened to electrolyte container. Follow these instructions closely. Scotch tape must also be removed from following places:

(1) Crankcase breathers on engine valve covers.

(2) Fuel tank breather pipe behind right seat in engine compartment.

(3) Torque converter breather pipe behind left seat in engine compartment.

(4) Transmission breather behind left seat in engine compartment.

TM 9-785
18-TON HIGH SPEED TRACTOR M4

  (5) Generator in engine compartment.
  (6) Distributor in engine compartment.
  (7) Air compressor air cleaner in engine compartment.
  (8) Breather on fan drive housing in engine compartment.
  (9) Breather on fan shaft housing in engine compartment.
  (10) Torque converter pump drive breather under rear floor plate.
  (11) Air exhaust at hand brake control.
  (12) Air exhaust at foot brake control.
  (13) Final drive breathers (one each side) under floor plates, driver's compartment.
  (14) Engine exhaust over engine compartment.
  (15) Air precleaner over engine compartment.
  (16) Air tank safety valve at rear of tractor.

  b. Remove the oil bath air cleaner cup to make sure it contains the correct amount and grade of oil (par. 77 b).

  c. Inspect the oil level in the engine crankcase, transmission case, final drive gear cases, winch gear housing, and level of fluid in torque converter reserve tank.

  d. Lubricate all other parts of the tractor (section VII).

  e. Check and fill fuel tank, if necessary, with the correct fuel. Special care must be taken to prevent the entrance of dirt or foreign materials while filling tank.

  f. Fill the cooling system with clean water that is free from lime or alkalines. In winter weather, use antifreeze solution (ethylene glycol type) in the cooling system. The solution should be tested daily and kept to the proper potency for the prevailing temperatures.

7. OPERATION OF NEW TRACTOR.

  a. Operate a new tractor with a light load during the first 60 hours. After the first ten hours of operation, stop the tractor, inspect for loose bolts and nuts, and check master clutch and steering brake adjustments (pars. 113 b and 123 b).

8. ENGINE STARTING INSTRUCTIONS.

  a. Figure 5 shows the operating controls and instruments. NOTE: Although "CHOKE" and "ENG. STOP" are shown mounted in instrument panel in the illustrations, later modifications made on tractor include mounting of these two controls on a bracket below panel.

  b. Each time before starting engine, perform the "Before-operation Service" as outlined in paragraph 26.

  c. Start Engine. If the engine has been idle and is cold, proceed as follows:

DRIVING CONTROLS AND OPERATION

```
A—DIMMER SWITCH
B—AIR BRAKE VALVE LEVER
C—STEERING LEVER LOCKS
D—MAIN LIGHT SWITCH
E—CHOKE
F—ENGINE OIL PRESSURE GAGE
G—ENGINE STOP
H—SPEEDOMETER
I—PANEL LAMPS
J—TORQUE CONVERTER PRESSURE
    GAGE
K—AMMETER
L—PRIMER PUMP
M—ENGINE TACHOMETER
N—TRANSMISSION OIL PRESSURE
    GAGE
O—FUEL GAGE
P—TRANSMISSION OIL TEMPERA-
    TURE GAGE
Q—HOUR METER
R—ELECTRIC TRAILER BRAKE LOAD
    CONTROL
S—WINDSHIELD WIPERS
T—PANEL LIGHT SWITCH
U—TRANSMISSION OIL TEMPERA-
    TURE GAGE
V—LOW AIR PRESSURE INDICATOR
W—TORQUE CONVERTER TEMPERA-
    TURE GAGE
X—REAR FLOODLIGHT SWITCH
Y—BLACKOUT LIGHT SWITCH
Z—ENGINE TEMPERATURE GAGE
AA—WINDSHIELD WIPER SWITCH
AB—SWITCH NOT USED
AC—STEERING LEVERS
AD—THROTTLE LEVER
AE—GEARSHIFT LEVER
AF—SIREN SWITCH
AG—CLUTCH PEDAL
AH—WINCH CLUTCH LEVER
AI—POWER TAKE-OFF LEVER
AJ—TRAILER BRAKE PEDAL
AK—ACCELERATOR PEDAL
AL—HULL DRAIN LEVER
AM—STARTER BUTTON SWITCH
                      RA PD 307564
```

Figure 5—Operator's Controls

(1) Make sure gearshift lever is in neutral position.

(2) Press in on button in center of knob and pull knob marked "ENG. STOP" all the way out. This closes the air valve in intake manifold and also closes ignition switch.

(3) Pull throttle control one-quarter of the way out.

(4) Press in on button in center of knob and pull choke control all the way out.

(5) Press on starter button.

(6) As soon as engine starts, push choke control in part way, then as engine warms up, continue to gradually push choke control in until it is against stop. Choke can also be regulated more accurately during engine warm-up period by turning knob of control. Turning knob clockwise tends to open choke valve; turning it counterclockwise tends to close choke valve. If this method is used to adjust choke, depress button in center of knob and push choke all the way in after engine has warmed up.

18-TON HIGH SPEED TRACTOR M4

(7) If engine has been stopped for only a short while and is still warm, *do not* use choke as it will result in flooding engine.

(8) When the atmospheric temperature is zero or below, it may be necessary to use the engine primer pump to aid starting. To start engine with aid of primer pump, perform steps (2), (3), and (4), then press on starter button. Pump plunger of primer pump two or three strokes (not over three) while engine is turning. This should deliver enough extra fuel to engine to start it. After engine starts, use choke control as explained in step (6) above.

(9) Check oil pressure. Oil pressure should read approximately 40 pounds on gage with engine running at operating speed. Pressure may be a little slow in coming up if oil is cold.

(10) Check engine temperature after engine has warmed up. Normal temperature is between 160° and 180°F. NOTE: *Always allow at least a five-minute warm-up period for engine before moving tractor.*

9. **STOPPING ENGINE.**

a. Close throttle and allow engine to slow down to idling speed. Then push engine stop knob all the way in against panel. Pushing in on knob turns off the ignition, and at the same time opens the air valve mounted between the two carburetors and connected by hose to the air cleaner outlet pipe. Opening this valve allows clean, fresh air to be drawn into the intake manifold and cylinders, and thus prevents fuel from being drawn in with the air. As no combustible mixture reaches the cylinders, the engine will stop. The tendency of a high-compression engine to continue to run through self-ignition after ignition is turned off is thereby overcome. Leave the engine stop knob in against panel until engine is to be started again.

10. **USE OF MASTER CLUTCH AND GEARSHIFT LEVER.**

a. The gearshift lever has five positions (fig. 6). Shift the lever into the "CREEPER" position if the slowest forward speed or greatest power is desired. Shift lever into "LOW" for intermediate forward speed or into "HIGH" for highest forward speed. The "ENGINE BRAKE" position is to be used only to hold tractor and load to a slow speed for descending steep grades. For backing tractor or load, shift lever into "REVERSE" position. Free wheeling is in effect when lever is in "LOW" position. A slotted plate with the lever positions marked on it and bolted below gearshift lever (fig. 121) will be found on all but some of the first tractors built. The gears may be shifted from "ENGINE BRAKE" or "LOW" to "HIGH" or vice versa while tractor is in motion.

DRIVING CONTROLS AND OPERATION

Gear	Gearshift Lever
Creeper range	Move lever to right and back from neutral position
Low range	Move lever forward from neutral position
High range	Move lever forward into "LOW," then to left and forward
Reverse	Move lever straight back from neutral position
Engine brake	Move lever from neutral position into "LOW," then to left and back

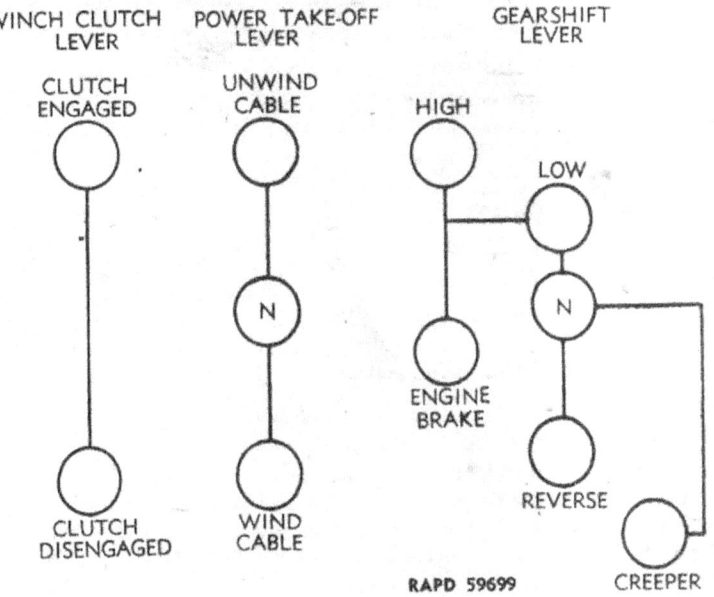

Figure 6—Diagram Showing Positions for Gearshift, Power Take-off and Winch Clutch Levers

b. To shift gears, push master clutch pedal all the way down and shift lever to position for desired speed. Have engine running at idling speed when shifting. Then let pedal up slowly until clutch is fully engaged, speed up engine, and tractor will begin to move. After tractor is in motion, use foot accelerator to maintain speed desired. To shift from a lower speed to a higher speed with tractor in motion, remove foot from accelerator, press clutch pedal all the way down, and shift gears. Let clutch pedal up quickly after lever has been shifted to desired position.

c. When pulling uphill in high range, and it becomes necessary to shift gears to low range, depress clutch pedal and pull gearshift lever out of the "HIGH" position. Gears will then automatically be in low range because the low gear remains in mesh when lever is shifted from "LOW" to "HIGH" position. This is made possible by the overrunning clutch in transmission. If for some reason the engine stops while pulling uphill in high gear, the gears will hold tractor from rolling back. However, if this happens, do not try to pull gearshift lever out of "HIGH" position until engine is again started. Depress clutch pedal and, with throttle lever in closed position, press starter button and start engine. Let up on clutch pedal, then pull back on gearshift lever with a firm pressure and increase engine speed slowly until lever slips out of "HIGH" position. Gears may then be shifted to a lower range in normal manner. Use same procedure if gears lock with lever in "ENGINE BRAKE" position.

d. It is desirable to operate the tractor in the speed range that will maintain a normal load on engine. Operating the tractor in a low gear with a light load and a resultant light load on engine, or operating in high gear with a heavy load and a resultant heavy load on engine, will cause excessive heating of the fluid in the torque converter. If operating in high gear and the load causes speed to drop to six or eight miles an hour, shift into a lower range. After tractor gains speed in the lower gear, to approximately 10 miles per hour, shift again into a higher gear. This will tend to maintain normal torque converter fluid temperature and a normal load on engine.

e. Do not slip master clutch. If it is necessary to keep engine running with tractor stopped, be sure gears are in neutral before taking foot from clutch pedal. CAUTION: *Do not ride clutch pedal while driving.*

f. The transmission may become locked should the engine stall when the tractor is attempting to negotiate a hill. The gears in the transmission can be unlocked when in the "ENGINE BRAKE" or "HIGH" position by pulling back on the brake levers (par. 11), accelerating the engine, and exerting pressure on the shifter levers at the same time. Should the transmission become locked in the "ENGINE BRAKE" position, it is not necessary to operate the clutch. If locked in the "HIGH" position, it may be necessary to depress the clutch quickly at the time pressure is being applied to the shifter levers.

11. STEERING LEVERS AND BRAKES.

a. Steering of the tractor is accomplished by means of the steering levers which operate brakes in the differential assembly. If it is desired to turn to the left, apply the left steering brake by pulling back on the left steering lever. This will retard the speed of the left track and increase the speed of the right track, causing the tractor to turn to the left. If right turn is desired, pull back on the right

DRIVING CONTROLS AND OPERATION

steering lever, which will slow down the right track and speed up the left, swinging the tractor to the right. Both tracks will turn unless throttle is closed and both steering levers are pulled back. The locks at top of the steering levers are for shifting brake lock ratchets into, or out of, operating position. With buttons turned down, the ratchet pawls will not engage ratchet quadrants. With buttons turned up, the ratchets are operative, and will hold levers at whatever point they are pulled back and hold brakes applied. Keep lock buttons turned down when traveling.

12. TOWING OF VEHICLE.

a. Towing Vehicle to Start Engine. In the event the engine cannot be started with the cranking motor, due to a defective cranking motor or discharged battery, and it is necessary to move or operate the tractor before repairs can be made to the cranking motor or battery charged, it can be started by pulling the tractor with another vehicle. Connect tow chains to clevises on front end of tractor and hook chains to vehicle to be used to pull the tractor. Pull engine stop button out and pull choke control out. Shift transmission gears of tractor into high gear and hold down on clutch pedal until towing vehicle gets tractor moving. Then let up on clutch pedal and engine will be turned for starting. If engine does not turn fast enough to start with transmission in high gear, shift lever into engine brake position (fig. 6). Engine will then be turned faster.

b. Towing Disabled Vehicle. If the tractor is to be towed to a repair shop due to an inoperative engine, torque converter, stripped gears, or other similar causes, have the gearshift lever in neutral position while towing. If tractor cannot be moved under its own power due to transmission or differential gears being locked or broken so that the tracks cannot turn, it will be necessary to remove the tracks to enable towing the tractor to where repairs can be made. Remove tracks as outlined in paragraph 133 c, then the tractor can be moved with the bogie and trailing idler wheels rolling on the ground. If this is done, care must be taken to avoid deep holes or large rocks that could cause damage to the bogie assemblies or bogie wheel tires. With tracks removed, the tractor brakes will be rendered useless, therefore a rigid towing bar should be used instead of a chain to prevent tractor from running into towing vehicle on downgrades. If possible haul tractor to repair shop on a truck.

13. USE OF TRAILER BRAKES.

a. Brakes on the trailer pulled by the tractor may be operated either by air or by electricity (depending on the type of brakes mounted on the trailing unit) since the tractor is equipped with complete controls and operating mechanism for both air brakes and

18-TON HIGH SPEED TRACTOR M4

electric brakes. Trailer brakes, of course, operate only on the load behind the tractor and are not effective in stopping the tractor itself. Refer to sections XXV and XXVI for more detailed information on trailer brake controls.

b. **Types of Trailer Brake Controls.** The trailer brake pedal to the left of the foot throttle (fig. 5), when pressed down, operates both the air and the electric brakes, while the hand control on the air valve at the left side of the dash is used for operating the air brakes by hand. Only one system is used at a time, depending on whether or not drawn vehicle is equipped with air or electric brakes.

c. **Connections.**

(1) Two air hose couplings are provided for connecting trailer air brake hose. The one on the left is marked "SERVICE" and the one on the right "EMERGENCY." The corresponding hose on the trailer should be connected to these couplings. When the tractor is operated without trailer, the cut-out cocks should be closed and the dummy couplings installed to keep dirt and dust out of system.

(2) The rear electric brake coupling socket on the tractor is wired so that when the plug from a trailer vehicle equipped with electric brakes is inserted, all electric brake and light apparatus will operate properly from tractor light switches and brake controls.

d. **Operation of Electric Brakes.** The brakes should be operated according to the requirements of the trailer. If a light braking effect is desired, the knob on the load controller at the right side of instrument panel should be set accordingly. Turning knob to left (counterclockwise) gives a light braking effect, and turning knob to right increases the braking power. The brake is applied by pressing down on the trailer brake pedal. Trailer stop light operates automatically. For further details of electric brakes, refer to section XXVI.

e. **Operation of Air Brakes.** The air brakes can be applied either by pressing down on the trailer brake pedal or by moving lever of hand-controlled air brake valve at left of dash to the right. The further the lever is moved or the further down the trailer brake pedal is pressed, the more air pressure is delivered to the brakes. Always use care not to slide wheels on trailer when stopping, unless emergency makes a sudden stop necessary. When it is necessary to stop the vehicle as quickly as possible, the operator can lock the wheels by moving the lever all the way to the right or pressing the trailer brake pedal all the way down until the stop is made. CAUTION: *When stopping tractor with towed vehicle attached, always apply the trailer brakes first.* After drag of trailer is felt, disengage master clutch and apply tractor brakes; jackknifing of tractor and trailer will thus be avoided. For further details of air brake system refer to section XXV.

DRIVING CONTROLS AND OPERATION

14. PARKING VEHICLE.

a. If tractor and trailer are to be parked, especially on a slope where there is a possibility of rolling, set the tractor steering brakes by pulling steering levers back as far as possible and engaging ratchet locks to lock them in that position. CAUTION: *Chock wheels on trailed vehicle if temperature is freezing or below instead of setting trailer brakes, as they are liable to freeze, especially if trailer has been pulled through water or slush.*

15. LIGHTING SYSTEM.

a. The lighting control system on this tractor is designed to operate both the lights on the trailed vehicle and those on the tractor. Proper connections for trailed vehicle are provided in the electric brake coupling socket at rear of tractor.

b. **Light Switches.** The main light switch on instrument panel has three positions for turning on the various lights. When the knob is pulled out to first stop, only the blackout lights are turned on. Turn on blackout driving light by pulling out knob marked "B-O DRIVE." For regular service lights, the small button latch on the top of the switch must be depressed and the knob pulled out to the second stop. To use stop lights only, for day driving, depress button latch and pull knob fully out. This light goes on automatically when both steering levers are pulled back.

c. The panel lights are turned on by pulling out on knob marked "PANEL LIGHT." These will not light unless headlights have first been turned on.

d. Turn on rear flood lamps by pulling out on button marked "REAR FLOOD LAMPS" at lower left side of speedometer. Headlights must be turned on before these lamps will light.

e. The service headlight dimmer switch is located on the left side of the dash above the siren switch and is operated by the left foot.

f. Turn off all lights by pushing knobs all the way in.

g. The siren switch is on the left side of the dash and is also operated by the left foot.

16. HOUR METER.

a. **Description.** This meter is electrically operated, and starts as soon as the engine ignition is turned on. The small hand at the top left of the dial will then start rotating. This hand is to indicate when the meter is operating.

b. **How to Read Hour Meter.**

(1) The three hands in the center of the dial record the number of hours the tractor has operated, and are of three different lengths. Total hours is determined by reading the number each hand has

TM 9-785
16

18-TON HIGH SPEED TRACTOR M4

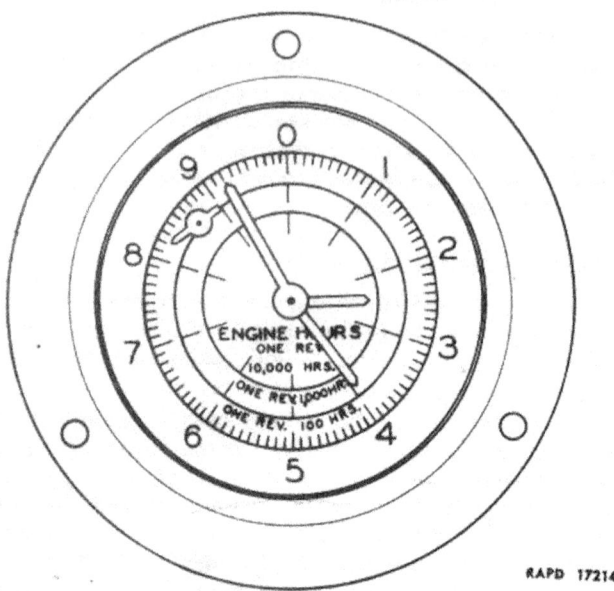

Figure 7—Hour Meter

passed, in the same manner as we look at a clock to see which number the hour hand, the minute hand, and the second hand have passed, and thereby tell the time.

(2) The shortest hand requires 1,000 hours of operation to pass each numeral, or 10,000 for a complete revolution. As illustrated in figure 7, it stands between 2 and 3, which indicates over 2,000 hours of operation. This indicates that the first numeral in the number of hours operated will be 2, followed by three other numerals indicated by the position of the other two hands, which in turn show how many hours more than 2,000 the tractor has operated.

(3) The medium-length hand requires 100 hours of operation to pass each numeral, or 1,000 hours for a complete revolution. In this illustration it stands between 3 and 4, which indicates over 300 hours which must be added to the 2,000. The second numeral then will be 3. Now observe the position of the longest hand to determine in the same manner how many hours must be added to the 2,300 indicated by the other two hands.

(4) This longest hand requires one hour of operation to pass each mark on the outer circle, 10 hours to pass each numeral, or 100 hours for a complete revolution. Here it stands on the second mark past 9, which indicates 92 hours to be added to the 2,300 so that the correct meter reading is 2,392.

DRIVING CONTROLS AND OPERATION

c. The small hand at the top left of dial may continue to rotate for approximately a minute after the engine is stopped.

17. **INSTRUMENTS** (fig. 96).

   a. **General.** The following instruments are provided to register the operation of various units of the tractor to enable the operator to tell by watching the instruments if these units are functioning properly. He should make a habit of glancing at these instruments often while driving, and if any of them register abnormally, stop the tractor and investigate before damage or breakdown results.

   b. **Ammeter.** The ammeter needle registers on the dial the amount of current being delivered to the battery by the generator when the engine and generator are running. If generator operation is normal, the needle swings toward the plus (+) side of the scale for a short while after starting engine, then needle returns nearly to zero after battery is again fully charged. When the engine is not running, current used by the electrical equipment (if turned on) discharges the battery, and the needle swings toward the minus (−) side of the scale.

   c. **Air Pressure Gage and Low Air Pressure Indicator.** The air pressure gage registers the pressure of the air in the air reservoir of the air brake control system. The pressure is regulated by the air pressure governor, which acts to maintain the pressure between 80 and 105 pounds. The red low air pressure indicator light comes on when the pressure drops below 60 pounds. CAUTION: *Do not use air brakes when red low air pressure indicator light is on.*

   d. **Engine Oil Pressure Gage.** The engine oil pressure gage is electrically operated and registers the pressure of the lubricating oil delivered to the engine. Normal oil pressure with engine at operating speed is approximately 40 pounds; however, there is no need for alarm unless pressure drops lower than 20 pounds.

   e. **Fuel Gage.** The fuel gage registers the level of the fuel in the fuel tank. Engine ignition switch must be closed for this gage to operate.

   f. **Speedometer.** The speedometer registers the speed of travel in miles per hour and also total miles traveled.

   g. **Tachometer.** The engine tachometer registers the speed of the engine in hundreds of revolutions per minute of the engine crankshaft.

   h. **Transmission Oil Pressure Gage.** This gage registers the pressure of the oil delivered to various units in the transmission, differential, and power take-off cases. Pressure will vary; however, there should be at least five pounds pressure maintained if tractor is being operated in the lower gears. Pressure gage may read zero while operating in high gear. If it does, there is no need for alarm unless

18-TON HIGH SPEED TRACTOR M4

transmission oil temperature gage also fails to register. If normal temperature is indicated, disregard a zero pressure reading.

i. **Transmission Oil Temperature Gage.** This gage registers the temperature of the oil in the transmission, differential, and power take-off cases. Temperatures as high as 250°F are allowable.

j. **Torque Converter Fluid Pressure Gage.** This gage registers the operating pressure of the fluid in the torque converter assembly. Normal operating pressure under load is from 40 to 50 pounds.

k. **Torque Converter Fluid Temperature Gage.** This gage registers the temperature of the fluid used in the torque converter. Operating tractor in low speed range under no load, or light load with high engine speed, will cause excessive heating of converter fluid. Shift gears into high speed range and use throttle to control tractor speed. Operating tractor in high speed range under heavy load, which reduces the tractor speed below six miles per hour, will also cause excessive heating of the converter fluid. Shift into low speed range. If temperature rises to above 220°F under normal operating conditions and with normal load, stop the tractor and inspect for cause of heating. In most cases, however, the torque converter fluid temperature will remain fairly low.

Section IV

OPERATION OF AUXILIARY EQUIPMENT

	Paragraph
Operation of winch and controls	18
Fire extinguisher	19
Tire inflation hose	20
Ammunition boxes and shell hoist	21

18. OPERATION OF WINCH AND CONTROLS.

a. To Attach Cable to Load. With gearshift lever in neutral position, depress master clutch pedal, push power take-off lever between steering brake levers forward (unwinding position) and push winch clutch lever on left fender of cab forward to engage jaw clutch on winch (fig. 6). It may be necessary to engage master clutch slightly to roll winch drive shaft, so jaw clutch may be engaged. Engage master clutch slowly and cable will unreel from drum. Unreel just enough cable so that hook can be attached to load, and disengage master clutch.

b. To Pull Load. After cable is attached, pull power take-off lever back until reverse notch is reached (winding position). Lock steering brakes with steering levers to hold tractor. Engage master clutch slowly until it is fully engaged, and speed up engine until winch will wind up cable and pull load. Speed of wind-up is controlled by accelerator. CAUTION: *Do not run engine faster than 1,400 revolutions per minute when pulling load with winch.* Always try to keep cable wound evenly and smoothly on drum. Stop winch by closing throttle and releasing master clutch. If load is to be moved a little at a time, cable will be held taut when master clutch is released by the automatic safety brake on the winch worm shaft. If cable is tight when pull is completed, move tractor backward to relieve strain so cable may be released. Wind cable up, release winch clutch, and move power take-off lever to neutral position. CAUTION: *Never pull out winch jaw clutch when winch is under load.*

19. FIRE EXTINGUISHER.

a. Description. A 4-pound CO_2 fire extinguisher (fig. 8) is part of the equipment of each tractor. It is carried in a bracket in rear seat compartment. The discharge of the contents, which is dry ice, is controlled by a trigger-operated valve and is aimed by a horn on the unit.

b. Operation. After removing the extinguisher from bracket, raise horn to a right angle with body of extinguisher, and point nozzle towards base of flame. Pull trigger (wire seal on trigger will break) and contents will be discharged. CAUTION: *Avoid contact of discharge with skin as frostbite will result from extended contact.* After

18-TON HIGH SPEED TRACTOR M4

Figure 8—Fire Extinguisher in Tractor

use, be sure extinguisher is sent back for recharging if enough of the contents has been used to lower the weight of the extinguisher more than six ounces below the weight stamped on valve boot or written on sticker on side of unit.

20. TIRE INFLATION HOSE.

a. A tire inflation hose is included in the equipment carried on tractor. For inflating tires requiring not more than 105 pounds air pressure, close cut-out cock in emergency trailer hose coupling on rear of tractor and remove dummy coupling (or disconnect trailer hose if a towed vehicle is connected to tractor). Connect tire inflation hose to coupling as shown in upper part of figure 9, then open cut-out cock and tire can be inflated. If tire requires more than 105 pounds air pressure, use the second connecting adapter furnished with hose and connect hose to the fitting on air supply valve (after removing cap from valve) instead of hose coupling as shown in lower part of figure 9. After connecting hose, turn lever of valve up. Air pressure governor is then cut off and pressure in air reservoir can be built up above 105 pounds and tire inflated. After tire is inflated and hose disconnected, turn valve lever back toward right side of tractor so air pressure governor can again control air pressure in reservoir.

TM 9-785
18-19

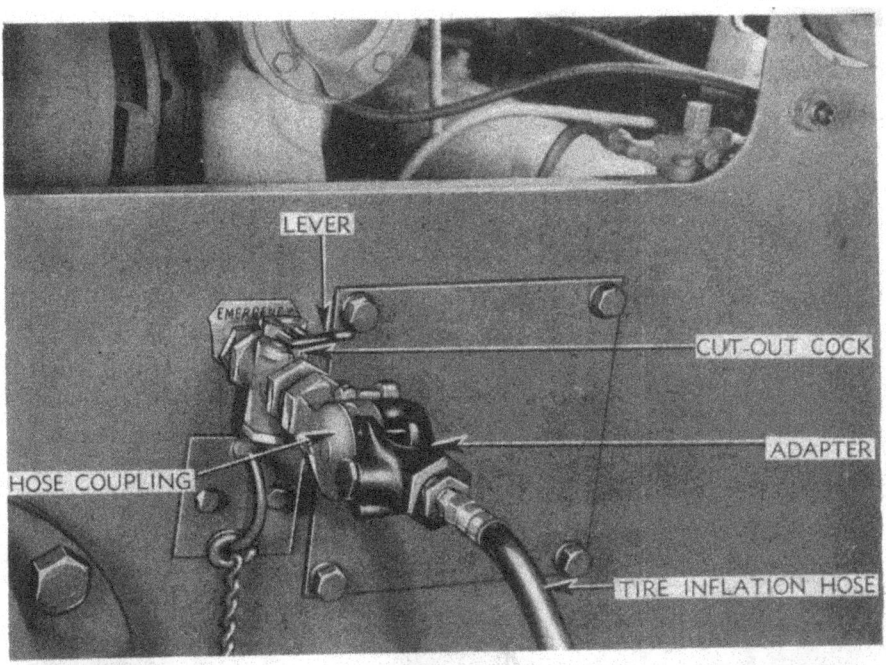

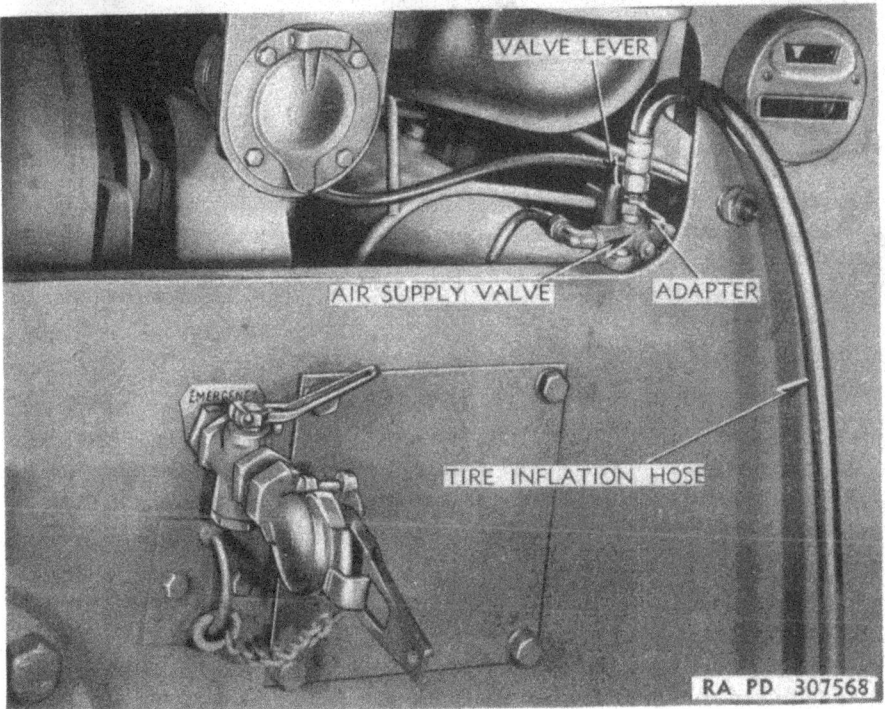

Figure 9—Tire Inflation Hose Connections

18-TON HIGH SPEED TRACTOR M4

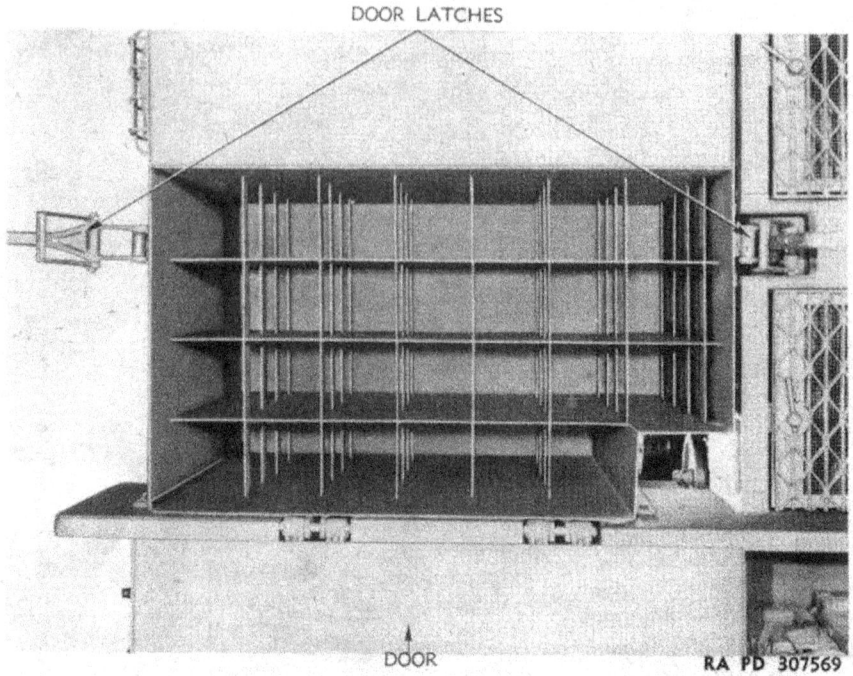

Figure 10—Shell Racks for 3-inch and 90-mm Ammunition

21. **AMMUNITION BOXES AND SHELL HOIST.**

   a. General. The tractor may be equipped with an ammunition box which provides stowage space for initial rounds of 90-mm or 3-inch antiaircraft ammunition (figs. 3 and 10), or with a cargo box for carrying initial rounds of ammunition for the 155-mm howitzer, 8-inch howitzer, or 240-mm gun (fig. 11). The two boxes are interchangeable on the tractor so that proper ammunition storage space can be provided for any one of the five guns mentioned above for which the tractor is adaptable as the prime mover.

   b. Preparation of cargo box and crane for loading 155-mm, 8-inch, or 240-mm ammunition. A special swinging crane with trolley hoist is provided with each cargo box for hoisting shells into box (fig. 11).

   (1) To prepare hoist for loading of shells into cargo box, remove cover and lower door of shell compartment, then unlock crane from top of box. Raise crane to position shown in figure 11 and insert lock pin into hole under crane shaft. Remove trolley stop pin at end of crane and place chain hoist on crane. Install trolley stop pin back in end of crane to prevent trolley from running off end. Crane is now ready for hoisting shells into box.

TM 9-785
21

OPERATION OF AUXILIARY EQUIPMENT

Figure 11—Crane and Hoist for Loading Shells

(2) Included with the cargo box are two pairs of shell racks and three pairs of hold-down plates. These are designed to properly space and hold the shells firmly in place. One pair of racks has round recesses in one side in which the bases of the 155-mm shells fit; when turned over, recesses in other side are the proper size for the bases of 8-inch howitzer shells. The second pair of racks has recesses in one side only into which fit the bases of the 240-mm howitzer shells. The pair of hold-down plates with the smallest holes is used when hauling 155-mm shells; the pair with the next larger size holes is for the 8-inch shells; and the pair with the largest holes is for the

27

18-TON HIGH SPEED TRACTOR M4

Figure 12—Arrangement for Stowage of Shell Racks and Hold-down Plates

240-mm shells. Cover plates and shell racks not used are racked into a compartment in bottom of shell compartment as explained in (3).

(3) Instructions for the stowage of shells is shown in figures 12 and 13 which illustrate with the use of 155-mm shells. The shell racks and hold-down plates for the 155-mm shells are used and the other pair of racks and two pair of cover plates must be placed in bottom of box in order and arrangement shown in figure 12. A hold-down plate is laid in first with bottom side up, a shell rack next with the top side up, then the second rack with bottom side up, and next the remaining three hold-down plates as shown. When stacked this way, the ribs and flanges on some of the plates fit in channels in others, and when all have been placed, the top of the stack of plates will be flush with, and form part of, the floor. Lay a pair of shell racks on floor next with side of racks with smallest holes up and fasten them to floor with four cap screws in each (fig. 13). Then bolt the eight hold-down clamps to sides of shell compartment, using lower rows of holes (use holes marked "A" when hauling 8-inch shells and holes marked "B" when hauling 240-mm shells).

(4) After shells have been placed in box, place hold-down plates over ends of shells. Turn hold-down blocks of clamps ¼ turn over plates and lock with lever. Adjust for proper tension of clamps with

TM 9-785
21

OPERATION OF AUXILIARY EQUIPMENT

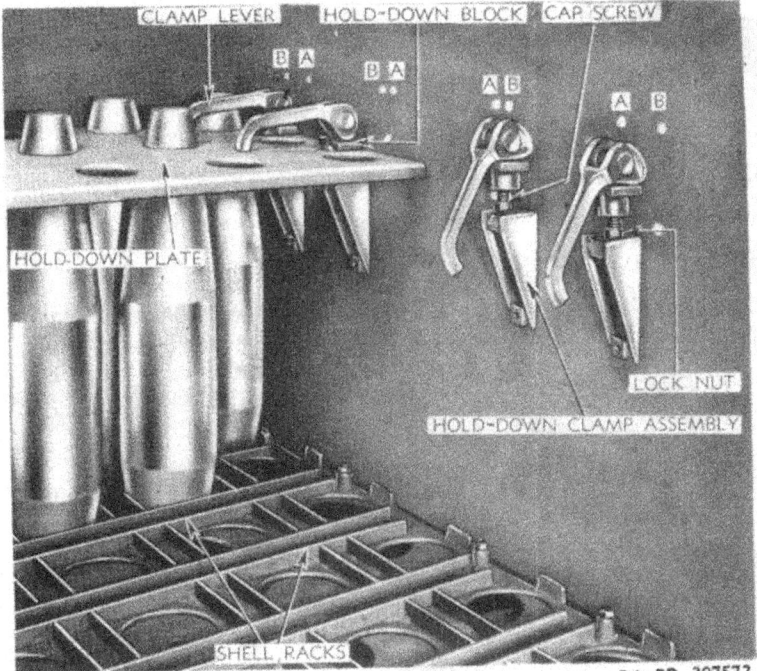

Figure 13—155-mm Shell Stowage Arrangement

cap screws in clamps (fig. 13). Tighten lock nuts when cap screws are adjusted properly.

(5) Remove chain hoist from crane and lay it in cargo compartment after shells have been loaded. Lower crane and secure end of crane with lock. Close door and install cover on box.

TM 9-785
22

18-TON HIGH SPEED TRACTOR M4

Section V

OPERATION UNDER UNUSUAL CONDITIONS

	Paragraph
Operating in extreme cold	22
Operating in water or mud	23
Operating in desert	24

22. OPERATING IN EXTREME COLD.

a. Antifreeze in Cooling Sytem. To protect the engine block and all other parts of the engine and the connections through which liquid from the cooling system flows, an adequate amount of suitable antifreeze liquid must be in the radiator at all times when temperature is at freezing point or below. The antifreeze used should have a higher boiling point than the normal operating temperature of the engine (160°F to 180°F). Always use antifreeze compound (ethylene glycol type) when available.

b. Lubricants. Lubricating oils and greases become thick when cold. Those issued for winter use should be thin enough to remain constantly fluid. Bearings will run dry and wear out if the oil or grease is too thick to flow freely between moving surfaces. Due to incomplete combustion in cold weather, gasoline is apt to drain into the crankcase and dilute the lubricating oil. Consequently, crankcase oil should be changed more frequently than in temperate climates.

c. Condensation. Moisture from the air has a tendency to condense on the inner surfaces of gasoline tanks and containers, and eventually freeze in gasoline lines and carburetors. This trouble can be alleviated to some extent by keeping containers as full as possible. Gasoline should be filtered through chamois to remove water before it is put into the tanks of vehicles.

d. Stopping Tractor. If tractor has been operating in mud, snow, or water, it is better when possible, to chock the wheels of the trailed unit at end of run when vehicle is parked than to set the brakes, as they will freeze and lock. Air tanks must be thawed frequently to drain condensed and frozen moisture. Rubber air hose becomes stiff and hard and requires constant inspections to determine any leakage. When stopping on snow, ice, or frozen ground, brakes should be applied gently. A sudden application of the brakes will cause accidents through slipping, jackknifing, or loss of control of the vehicle.

e. Batteries. Starting batteries are much less efficient when the weather is cold and engine is hard to turn over. To assure maximum performance from batteries, all battery connections should be cleaned frequently and voltage checked to assure that batteries are properly

TM 9-785
22

OPERATION UNDER UNUSUAL CONDITIONS

charged. The liquid in batteries should cover the surface of plates at all times. Water should be added to batteries just prior to running engine, so that there will be complete circulation and mixing of all liquid in batteries. Water added to batteries which are to remain idle for some time does not circulate and will freeze in very cold weather.

f. **Cold Weather Starting.** Starting cold engines is always difficult. The most satisfactory method of starting engines which have been exposed to extremely low temperature is to cover the entire vehicle with a tarpaulin and place either one or two jet type burners under the vehicle (refer to subpar. g). This method of heating warms the entire unit, insuring easier starting and complete flexibility of

Figure 14—Heater Installed at Rear of Tractor

all moving parts when unit is put into operation. One of the most important factors in starting engines which have been exposed to very low temperatures is the storage batteries. A battery exposed to —40°F, even though fully charged at time of exposure, will have no cranking energy and must be warmed before it will offer any cranking ability. The use of stand-by heat such as described above warms the batteries as well as the rest of the unit and makes easy starting possible.

g. **Use of Heaters and Covers in Cold Weather.**

(1) GENERAL. Fuel lines, valves, and connections are provided

TM 9-785
22

18-TON HIGH SPEED TRACTOR M4

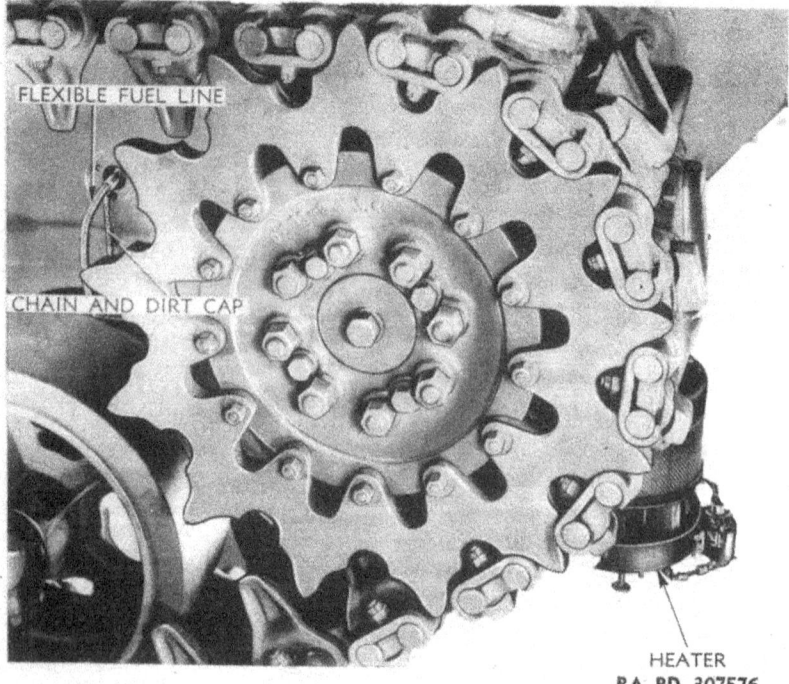

Figure 15—Heater and Connection at Front and Side of Tractor

OPERATION UNDER UNUSUAL CONDITIONS

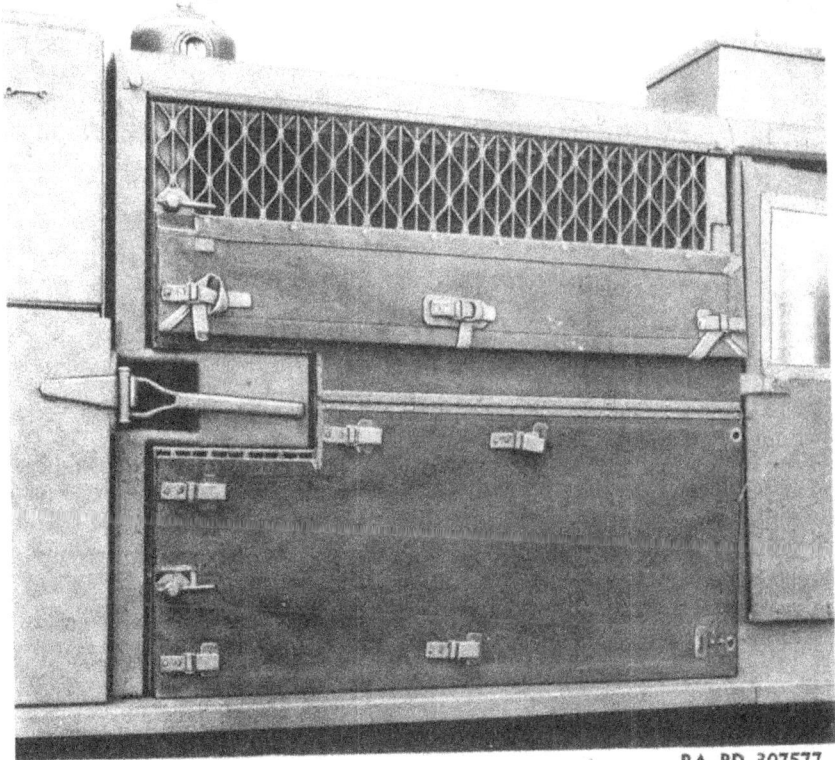

Figure 16—Covers for Cold Weather Operation

18-TON HIGH SPEED TRACTOR M4

on tractor for use of heaters in extremely cold weather to keep tractor at temperature at which engine can be readily started and moving parts and lubricants will be in a flexible condition. This is done by covering tractor with a tarpaulin or using special canvas covers designed to cover radiator and engine grilles and lower part of tractor, and by using either one or two heaters inside the tarpaulin. These heaters use gasoline from the tractor fuel tank. The special radiator cover shown in upper part of figure 16 has hinged sections to allow one-half, three-quarters, or all of radiator to be covered to help maintain correct engine operating temperature.

(2) How to Connect Heaters. Remove the two upper cap screws, loosen bottom cap screw, and swing down or remove cover at rear of hull, as shown in figure 14. Remove dirt cap from end of flexible fuel line on heater and connect end of line to the connection in rear of tractor. Open valve by turning it ¼ turn (crosswise of tractor). Fuel will now flow from tractor fuel tank to heater. Hang heater on rear of tractor so the base of heater is off the ground, then follow directions on heater for operation. A second heater can be used by connecting it to another heater connection at right side of tractor as shown in upper part of figure 15 and hanging it on front of tractor as shown in lower part of figure 15. A clip on which to hang the heater is furnished with heater. Install clip as shown in lower part of figure 15 on one of the final drive cover plate cap screws. Lift front seat plate and turn valve lever under seat ¼ turn (crosswise of tractor) to supply fuel to this heater. After heaters are burning properly, cover tractor with tarpaulin or with special winter covers as shown in figure 16. CAUTION: *Constant watch must be kept to guard against fire hazards, and to make sure heaters continue to burn Close both valves in tractor before disconnecting heaters (valve lever at front of tractor should point forward and rear valve lever should point down when closed). Be sure to replace covers on tractor, and dirt caps on ends of heater lines.*

23. OPERATING IN WATER OR MUD.

a. Make sure hull drain valve is closed and all drain plugs in bottom of hull are installed before entering water or operating in mud. See that covers are in place and secure on gun or ring in top of cab and on ammunition box and trailed unit. As soon as practical after emerging from water, depending on tactical situation, inspect for presence of water in hull and open hull drain or remove appropriate drain plugs in hull to drain water from hull, especially in freezing weather.

24. OPERATING IN DESERT.

a. General. Some of the more frequent difficulties experienced with engines in the desert are given below. By foresight and careful

OPERATION UNDER UNUSUAL CONDITIONS

driver and mechanic training many of these difficulties can be prevented or overcome.

b. Overheating. Overheating is a general complaint. It is particularly apt to happen when driving in the heat of the day with a tail wind, and on soft ground in low gear. Vehicle should be stopped, if possible, and allowed to cool. Fan belts should be kept in proper adjustment and examined frequently and replaced if they show wear, otherwise, they may break and, if not discovered at once, the resulting overheating may crack the engine block. Keep radiator clear of leaves, trash, etc. so air flow will not be restricted. Be sure water pump is in good operating condition, also that thermostat, if one is installed, opens fully, and water passages are free of restrictions.

c. Excessive Wear Due to Sand. Sand is present in the air at all times and scored cylinders, badly worn main bearings, crankshaft journals, etc. result from sand getting into working parts. Air cleaners must be inspected and cleaned much more frequently when vehicle is being operated under desert conditions, or in sandy soil, than when traveling over normal terrain. The elements in oil filters must be changed and fuel filter cleaned much oftener than usual. Sand may enter the crankcase and fuel lines when replenishing the oil or refueling unless great care is used during these operations. The driver must make an invariable habit of wiping all sand off the filler caps, before removing, and off the spouts of oil and gasoline containers, before they are used. Sand and dust will cause failure in operation of instruments in panel if these are not carefully sealed. Scotch tape may be used for this purpose.

d. Electrical Difficulties. The constant shock and vibration caused by passage over rough ground frequently causes cable clips to shake loose and cables to be broken or shorted. Frequent inspections of cable clips should be made and spring washers inserted under the nuts if possible. Voltage control units may cause trouble because of a broken wire in shunt winding or sticking of regulator points. Drivers must be trained to watch the ammeter as carefully as any other instrument since over-charging, even when not sufficient to buckle the plates, always results in loss of battery water, most difficult to obtain in the desert. Drivers must be warned that the high salinity of water issued for drinking and for radiators forbids its use in batteries.

TM 9-785
25

18-TON HIGH SPEED TRACTOR M4

Section VI

FIRST ECHELON PREVENTIVE MAINTENANCE

	Paragraph
Purpose	25
Before-operation service	26
During-operation service	27
At-halt service	28
After-operation and weekly service	29

25. PURPOSE.

a. To insure mechanical efficiency, it is necessary that the vehicle be systematically inspected at intervals during the day, and each week, in order that scheduled maintenance services be performed, and that defects may be discovered and corrected before they result in serious damage or failure. The services set forth in this section are those performed by driver or crew before operation, during operation, at halt, and after operation (and weekly).

b. Driver Preventive Maintenance Services are outlined for all vehicles in general on the back of "Driver's Trip Ticket and Preventive Maintenance Service Record," W. D. Form No. 48. However, in this section, certain procedures that do not apply to this vehicle are deleted, and in some cases there are deviations from the numerical sequence of the item numbers in order to best conserve driver's time and effort in the performance of the services.

c. The general inspection of each item applies also to any supporting member or connection, and usually includes a check to see whether or not the item is in good condition, correctly assembled, secure, or excessively worn.

d. The inspection for "good condition" is usually an external visual inspection to determine whether or not the unit is damaged beyond safe or serviceable limits. The term "good condition" is explained further by the following: not bent or twisted, not chafed or burned, not broken or cracked, not bare or frayed, not dented or collapsed, not torn or cut.

e. The inspection of a unit to see that it is "correctly assembled" is usually an external visual inspection to see whether or not it is in its normal assembled position in the vehicle.

f. The check of a unit to determine if it is "secure" is usually an external visual inspection, a hand-feel, or a pry-bar check for looseness in the unit. Such an inspection should include any brackets, lock washers, lock nuts, locking wires, or cotter pins used in the assembly.

g. "Excessively worn" will be understood to mean worn close to, or beyond, serviceable limits and likely to result in a failure if not replaced before the next scheduled inspection.

FIRST ECHELON PREVENTIVE MAINTENANCE

h. Any defects or unsatisfactory operating characteristics beyond the scope of first echelon to correct must be reported at the earliest opportunity to the designated individual in authority.

26. BEFORE-OPERATION SERVICE.

a. This service will not be entirely omitted, even in extreme tactical situations. When thoroughly trained, driver will be able to quickly determine the condition of the vehicle.

b. **Procedures.** Before-operation Service consists of inspecting items listed below according to the procedure described, and correcting or reporting any deficiencies. Upon completion of the service, results should be reported promptly to the designated individual in authority.

(1) ITEM 1, TAMPERING AND DAMAGE. Look for any injury to vehicle in general and to its accessories or equipment caused by tampering, sabotage, collision, falling debris or shell fire since parking vehicle. Look in engine compartment, both front and rear seat compartments, and in hull for signs of tampering or sabotage such as loosened or damaged accessories, loose fuel, oil, or water lines, disconnected wiring or linkage. If electrical wiring or accessories appear to be wet, dry them to facilitate starting.

(2) ITEM 2, FIRE EXTINGUISHER. Inspect for good condition, firm mounting and intact seal.

(3) ITEM 3, FUEL, OIL AND WATER. Note amount of fuel in tanks and spare cans. Add fuel if necessary. Check crankcase oil level. Add oil, if necessary, to bring level to "FULL" mark. Check level and condition of coolant and in cold weather when antifreeze is used have value of antifreeze checked if loss of coolant has been considerable.

(4) ITEM 4, ACCESSORIES AND DRIVES. Examine all accessories such as carburetors, generator, regulator, starter, fuel pump, fuel filter, fan and water pump for loose connections and mountings. Inspect carburetors, fuel pump, fuel filter, and water pump for leaks.

(5) ITEM 5, AIR BRAKE TANKS. Inspect trailer air brake valve assemblies for leaks with trailer brakes applied, and released. See that dummy couplings are in place on hose couplings if no towed vehicle is connected. Check hose connections if hooked up to towed vehicle. Look at air brake reservoir tank to see if it is mounted securely and not damaged. Drain water (condensation) from air reservoir. **CAUTION:** *Close pet cocks.* **NOTE:** *Complete draining is obtained only with air pressure in tank.* Drain after starting engine if no pressure is left from previous operations.

(6) ITEM 6, LEAKS, GENERAL. Look under vehicle and on ground for any indications of fuel, oil, water, gear oil, or fluid leaks. Look in bottom of hull for evidence of leaks. Drain hull by removing appropriate plugs if necessary.

18-TON HIGH SPEED TRACTOR M4

(7) ITEM 7, ENGINE WARM-UP. Start engine. Observe any tendency toward hard cranking, low cranking speed, or improper or noisy engaging or disengaging of starter drive. Set throttle to 900 revolutions per minute (tachometer reading) and allow engine to warm up while proceeding with the following Before-operation Service. During engine warm-up listen for unusual noises or misfiring; observe instrument readings, engine temperature rise, and engine performance. CAUTION: *If engine oil pressure does not reach 5 to 10 pounds at idling speed immediately after starting, stop engine and report.*

(8) ITEM 8, CHOKE. While starting engine, test for operation of choke and examine linkage and connections for looseness. As engine warms up, gradually push in choke button to prevent over-choking engine oil dilution.

(9) ITEM 9, INSTRUMENTS. Observe instruments during warm-up as follows:

(a) Engine Oil Pressure Gage. Normal reading of gage is 30 to 40 pounds at 900 revolutions per minute or fast idle speed. Pressure at slow idle is 10 to 15 pounds. CAUTION: *If gage does not indicate normal pressure, stop engine immediately and investigate cause. Pressure may be slightly higher before engine warms up.*

(b) Ammeter. After engine starts and is running at fast idle, the ammeter should show a high positive (+) charge rate for a short period until generator restores battery current used in starting. After this period, ammeter should register zero or slight positive charge with lights and accessories turned off.

(c) Fuel Gage. Gage must register approximate amount of fuel in tank.

(d) Engine Temperature Gage. Normal engine operating temperature is 160°F to 180°F. Engine temperature should increase gradually during warm-up period.

(e) Tachometer. Engine revolutions per minute in hundreds should be indicated without undue noise, or fluctuation of needle.

(f) Air Pressure Gage (and low air pressure indicator light). If vehicle with air brakes is to be towed, do not move tractor until air pressure gage shows 85 pounds pressure. Low air pressure indicator light remains "ON" until pressure is above 50 to 60 pounds. If 85 pounds pressure is not reached within reasonable engine warm-up time, investigate for faulty compressor operation, open valves or leaks. Maximum governed pressure is 105 pounds.

(g) Torque Converter Fluid Pressure Gage. Gage should show reading slightly above zero with engine operating and vehicle standing.

(h) Torque Converter Fluid Temperature Gage. Maximum allowable temperature is 220°F.

(i) Transmission Oil Pressure Gage. Gage should show a slight pressure with vehicle standing.

TM 9-785
26-27

FIRST ECHELON PREVENTIVE MAINTENANCE

(*j*) *Transmission Oil Temperature Gage.* This gage may register zero before tractor has been in operation. Oil temperature will rise during operation. Maximum allowable temperature is 250°F.

(10) ITEM 10, SIREN AND WINDSHIELD WIPERS.

(*a*) *Siren.* Tactical situation permitting, test operation of siren.

(*b*) *Windshield Wipers.* Operate windshield wipers; observe for blade contact and full stroke.

(11) ITEM 11, GLASS. Inspect for damaged frames and discolored glass. Clean windshield glass and celluloid in side curtains.

(12) ITEM 12, LAMPS AND REFLECTORS. Tactical situation permitting, turn on switches and see that all lamps light. See that all lamps are secure and lamp lenses and reflectors are clean and not broken.

(13) ITEM 15, SPRINGS AND SUSPENSIONS. See that track tension is correct (three-fourths inch sag between track support rollers). Look for loose or excessively worn blocks, worn connectors and guides, and bottomed wedges. Dead blocks will be identified by one end of shoe being higher than other. Examine mounting bolts on drive sprockets, trailing idlers, bogie wheels, and track support rollers. If two or more coils of volute springs are resting on seat the springs will be considered to have taken a permanent set and require replacement. Clean debris, stones, or excessive dirt from volute springs and track before moving vehicle.

(14) ITEM 18, TOWING CONNECTIONS. See that connections are not lost or damaged.

(15) ITEM 19, BODY, LOAD, AND TARPAULINS. Inspect to see that cargo is secure and covers are in good condition.

(16) ITEM 20, DECONTAMINATOR. Inspect decontaminator for full charge (shake), closed valve, and secure mounting.

(17) ITEM 21, TOOLS AND EQUIPMENT. Inspect for presence and proper stowage (check list, pars. 36 and 37).

(18) ITEM 22, ENGINE OPERATION. Engine should idle smoothly at operating temperature with choke button fully depressed. Accelerate and decelerate engine and listen for unusual noises that may indicate loose, worn, damaged or inadequately lubricated engine parts or accessories. Note if unusual amount of black smoke issues from exhaust.

(19) ITEM 23, DRIVER'S PERMIT AND FORM NO. 26. Make sure that Driver's Permit, Accident Report, Form No. 26, vehicle manual, and parts list are in the vehicle, are legible and safely stowed.

(20) ITEM 25, DURING-OPERATION SERVICE. This service must start immediately after vehicle is put in motion.

27. DURING-OPERATION SERVICE.

a. With vehicle in motion, listen for rattles, squeals or hums that may indicate trouble. Watch for steam from radiator and smoke

TM 9-785

18-TON HIGH SPEED TRACTOR M4

from any part of vehicle. Learn to detect odor of overheated carburetor, driving units, exhaust gas, or other signs of trouble. Observe while operating if brakes operate properly and gears shift as they should. Watch instruments constantly for possible trouble in systems to which instruments apply.

b. **Procedures.** During-operation Service consists of observing items listed below, according to the procedures. Note minor deficiencies to be corrected or reported at the earliest opportunity, usually at next scheduled halt. Stop vehicle if serious trouble develops.

(1) ITEM 26, STEERING BRAKES. Before starting vehicle test operation of steering brakes. Top of lever should move back 6 inches before brake engagement begins, with complete engagement taking place at or just ahead of vertical position. Both levers should pull back evenly. During operation, test right and left steering brake. On level stretches observe whether or not vehicle travels straight ahead with both levers released (right or left pull indicates faulty steering brake adjustment or misalinement of track). Test for proper engagement of lever locks.

(2) ITEM 28, CLUTCH. Clutch pedal should have one and one-half inches free travel before disengagement of clutch begins. Clutch should not chatter, squeal or slip. CAUTION: *Do not ride clutch.*

(3) ITEM 29, TRANSMISSION. Gears must shift smoothly, operate normally, and not slip out of engagement during operation.

(4) ITEM 31, ENGINE CONTROLS. Be alert for any deficiency in engine performance such as lack of usual power, misfiring, unusual noises, stalling, overheating, or unusual exhaust smoke. Note whether or not engine responds to controls and whether or not controls seem to be in proper adjustment.

(5) ITEM 32, INSTRUMENTS. Observe all instruments to see if systems to which they apply are functioning normally.

(a) Engine Temperature Gage. Normal temperature is 160°F to 180°F, except under unusual conditions.

(b) Engine Oil Pressure Gage. Normal pressure is 10 to 15 pounds while idling and 30 to 40 pounds at vehicle operating speeds. Lack of pressure requires immediate stopping of engine.

(c) Ammeter. Ammeter should indicate zero or positive (+) reading during operation. Negative reading (−) may indicate faulty generator, regulator, or other trouble.

(d) Fuel Gage. Gage should indicate approximate amount of fuel in tank.

(e) Speedometer. Speedometer should indicate vehicle speed without noise or fluctuation.

(f) Odometer. Odometer should record accumulating trip and total mileage.

FIRST ECHELON PREVENTIVE MAINTENANCE

(g) Tachometer. Tachometer should register engine revolutions per minute and revolutions per minute count.

(h) Air Pressure Gage. Gage should register between 80 and 105 pounds during operation. Low air pressure indicator light (if "ON") indicates air pressure is too low for operation of trailer brakes (50 to 60 pounds or less). Maximum governed pressure is 105 pounds.

(i) Transmission Oil Pressure Gage. Gage must show at least 5 pounds pressure while tractor is in operation.

(j) Transmission Oil Temperature Gage. Maximum allowable temperature is 250°F.

(k) Torque Converter Fluid Pressure Gage. Normal operating pressure under load is 40 to 50 pounds.

(l) Torque Converter Fluid Temperature Gage. If temperature rises above 220°F under normal operating conditions and with normal towed load, stop tractor and inspect for cause.

(m) Hour Meter. Meter should be registering during time engine is operating. Observe whether or not small indicator hand is rotating for check on operation.

(6) ITEM 34, RUNNING GEAR. Listen for any unusual noises from track or suspension parts that may indicate loose or damaged parts.

(7) ITEM 35, BODY AND TRAILER. Be alert for abnormal sagging or tilting of tractor; loose tarpaulins; unusual weaving of towed loads; loose bolts, hardware, gun mounts, or equipment.

(8) ITEM 36, GUNS. Listen for unusual noises indicating loose gun mount.

28. AT-HALT SERVICE.

a. At-halt Service may be regarded as minimum battle maintenance and should be performed under all tactical conditions even though more extensive maintenance service must be slighted, or omitted altogether.

b. **Procedure.** At-halt Service consists of investigating any deficiencies noted during operation, inspecting items listed below according to the procedure following the items, and correcting any deficiency found. Deficiencies not corrected should be reported promptly.

(1) ITEM 38, FUEL, OIL, AND WATER. Replenish as required. Have hydrometer check made of coolant if loss of coolant has been considerable during period when antifreeze is used.

(2) ITEM 39, TEMPERATURES. Hand feel the following for overheating: final drive cases, sprocket hubs, track support rollers, bogie wheels, and trailing idler hubs.

(3) ITEM 42, SPRINGS AND SUSPENSIONS. Inspect for broken springs. Inspect sprocket bolts, also bolts holding bogie assemblies,

18-TON HIGH SPEED TRACTOR M4

track support rollers, and trailing idlers to side of hull. Tighten any loose bolts. Make sure plugs in ends of bogie wheel, trailing idler, and track support roller shafts are tight.

(4) ITEM 45, TRACK. Inspect for dead or damaged track blocks and lost or bottomed wedges. Observe if track has required three-fourths inch sag between track support rollers. Clean out stones or debris from tracks and volute springs.

(5) ITEM 46, LEAKS, GENERAL. Look under vehicle and in hull compartments for indications of fuel, oil, water, or grease leaks.

(6) ITEM 47, ACCESSORIES AND BELTS. Examine accessories for loose mounting bolts, damage, or incorrect alinement. Observe condition of generator, air compressor, and fan drive belts. Inspect belts for three-fourths to one-inch deflection.

(7) ITEM 48, AIR CLEANERS. Cleaners must be secure and air passages clean. When operating under extremely dusty and sandy conditions, inspect air cleaners, breather caps, and breather pipe openings frequently for condition to deliver clean air properly. When inspection of engine air precleaner shows dirt level is ½ way up on the indicator glass, the precleaner must be emptied. If it is necessary to empty precleaner more than twice in one day, the oil bath air cleaner cup should be cleaned and refilled every second time the precleaner is emptied. If engine air cleaner requires frequent service, air compressor air filter or oil bath air cleaner, as equipped, should likewise be serviced at same interval.

(8) ITEM 50, TOWING CONNECTIONS. All towing connections must be securely fastened and locked. Observe for air leaks if using air brake hook-up.

(9) ITEM 51, BODY, LOAD AND TARPAULIN. Inspect ammunition compartment for secure fastening of ammunition. Make sure that tarpaulins are properly secured.

(10) ITEM 52, APPEARANCE AND GLASS. Clean windshield windows, celluloid in side curtains, and lamp lenses, and inspect vehicle and towed vehicle for damage.

29. AFTER-OPERATION AND WEEKLY SERVICE.

a. After-operation Service is particularly important because at this time the driver inspects his vehicle to detect deficiencies developed and correct those he is permitted to handle. He should report promptly the results of his inspection. If this schedule is performed thoroughly, vehicle should be ready to move again on a moment's notice. The Before-operation Service, with a few exceptions, is then necessary only to ascertain whether or not vehicle is in same condition in which it was left. After-operation Service should never be entirely omitted even in extreme tactical situations but may be reduced to the bare fundamental service outlined for the At-halt Service, if necessary.

FIRST ECHELON PREVENTIVE MAINTENANCE

b. **Procedure.** When performing the After-operation Service the driver must remember and consider any irregularity noticed during the day in the Before-operation, During-operation and At-halt Services. The After-operation Service consists of inspecting or testing the following units and correcting or reporting any deficiency. Those items of the After-operation Service that are marked by an asterisk (*) require additional weekly service procedure and are indicated in step *(b)* of each applicable item.

(1) ITEM 56, INSTRUMENTS. Before stopping engine inspect instruments for secure mounting, proper connections, or damage.

(2) ITEM 54, FUEL, OIL AND WATER. Check coolant and oil levels, add as needed. Fill fuel tanks, also refill spare cans.

(3) ITEM 55, ENGINE OPERATION. Test for smooth idling of engine without stalling. Accelerate and decelerate engine, noting if engine misses, or backfires, or whether or not unusual noises or vibrations indicate worn parts, loose mountings, incorrect fuel mixture, or faulty ignition. Investigate any unsatisfactory engine operating characteristics noted during operation.

(4) ITEM 57, SIREN AND WINDSHIELD WIPERS. Inspect siren for secure mounting and proper connections. Inspect wiper motor for security and good condition and wiper arms and blades for smooth operation and full stroke.

(5) ITEM 58, GLASS. Clean windshield and celluloid windows; report if damaged.

(6) ITEM 59, LAMPS AND REFLECTORS. If tactical situation permits, observe whether or not lamps light when switched "ON" and go out when switched "OFF". Inspect all lenses for dirt or damage; clean if necessary.

(7) ITEM 60, FIRE EXTINGUISHER. Inspect extinguisher for tight mounting, unbroken seal, and leakage at valve. If extinguisher has been used, report it for refill or replacement.

(8) ITEM 61, DECONTAMINATOR. Inspect to see whether or not decontaminator is full (test by shaking) and make sure it is mounted securely. Decontaminators require replacement every three months. Refer to tag for date of last recharge.

(9) ITEM 62, *BATTERY.

(a) Inspect battery for cleanliness and good condition, secure mounting and connections, and proper electrolyte level. See that vent caps are clean and secure.

(b) Weekly. Clean dirt from top of battery and remove caps. Add clean (drinkable) water to bring level $3/8$ inch above separator plates, if necessary. CAUTION: *Do not overfill. If terminals are corroded, clean and grease posts and terminals lightly, then carefully tighten terminals and mounting bolts.*

18-TON HIGH SPEED TRACTOR M4

(10) Item 63, *Accessories and Belts.

(a) Inspect carburetors, generator, regulator, starter, air compressor, fan and water pump for loose connections or mountings. Inspect carburetor and water pump for leaks. Examine belts for three-fourths to one-inch deflection.

(b) *Weekly.* Tighten, if necessary, all bolts and mountings on accessories such as carburetor, air cleaner, air compressor, starter, fan and water pump. Look for loose connections or leaks.

(11) Item 64, *Electric Wiring.

(a) Inspect ignition wiring and spark plugs for secure connections, cleanliness, and good condition.

(b) *Weekly.* Inspect all accessible low voltage wiring, make sure connections are tight and wires are in good condition.

(12) Item 65, *Air Cleaners and Breather Caps.

(a) Examine oil in air cleaners for correct level and excessive dirt. If oil is excessively dirty, clean and refill with fresh oil. Empty dirt from air precleaner when performing this operation. At all times and especially when operating in dusty or sandy territory close watch should be kept on glass indicator in engine air precleaner. Dirt accumulation in precleaner should not be permitted to rise above ½ way point on glass before emptying. If necessary to empty precleaner more than twice in one day, the engine air oil bath cleaner should be cleaned out thoroughly every second time engine air precleaner is emptied and the oil reservoir cleaned and replenished with new, fresh oil.

(b) *Weekly.* Clean and service engine air cleaner breather pipe, air cleaner, and air compressor air filter (or oil bath air cleaner, as equipped).

(13) Item 66, *Fuel Filter.

(a) Make sure filter is in good condition and does not leak.

(b) *Weekly.* Remove drain plug from filter, allow water and sediment to drain out of bowl, and replace plug.

(14) Item 67, Engine Controls. Make sure all controls work freely and are not disconnected, worn, or damaged.

(15) Item 68, *Tracks.

(a) Inspect tracks for dead or damaged track blocks and lost or bottomed wedges. Inspect track for three-fourths-inch sag between track support rollers. Clean out stones or debris from tracks and springs.

(b) *Weekly.* Tighten wedges.

(16) Item 69, *Springs and Suspensions.

(a) Thoroughly inspect drive sprocket, bogie, track support roller, and trailing idler assemblies. Make sure plugs in shafts are tight and tracks in alinement. Inspect springs for breakage.

FIRST ECHELON PREVENTIVE MAINTENANCE

(b) *Weekly.* Tighten or replace loose or missing bolts.

(17) ITEM 71, PROPELLER SHAFTS. Inspect for loose connections, lubricant leaks, or damage.

(18) ITEM 72, *VENTS.

(a) Examine following vents for good condition and clear opening: transmission and differential vent, torque converter vent, fuel tank vent, final drive vents. Remove any obstructions. Examine torque converter pump compartment breather and filler cap, crankcase breathers in filler caps in rocker arm covers, and fan housing and fan shaft housing breather and filler caps for good condition and unobstructed openings.

(b) *Weekly.* Remove and clean breathers and vents and make sure vent lines are open.

(19) ITEM 73, LEAKS, GENERAL. Look beneath vehicle and in hull compartments for indications of fuel, oil, and water leaks. Examine transmission, controlled differential and final drive, and torque converter for leaks. Inspect winch for oil leaks. Trace all leaks to source and correct or report.

(20) ITEM 74, GEAR OIL LEVELS. Check lubricant levels as follows: Differential and transmission on bayonet gage—fill with seasonal grade of engine oil to "FULL" mark. Final drives (two)—fill to level of filler plug with seasonal grade of engine oil. Torque converter pump drive housing (1 point)—fill, if necessary, with engine oil to level plug. Torque converter—open drain cock after fluid has had time to settle (5 minutes) and drain, not to exceed ½ cup fluid. CAUTION: *Perform this draining weekly except in freezing weather, then it should be done daily.* Check fluid level in fluid reservoir with bayonet gage. Add fluid to raise level to "FULL" mark, if necessary. Fan drive and gear housings should be filled to level plug.

(21) ITEM 75, *AIR BRAKE TANK AND CONNECTIONS.

(a) Listen at several points about tractor and towed vehicle for indications of air leaks. Drain water (condensation) from air reservoir and close valve after draining.

(b) *Weekly.* Drain condensation from tanks. Inspect system for good condition.

(22) ITEM 77, *TOWING CONNECTIONS. Make sure towing clevises are not lost or damaged. Be sure load is secure and safety latches locked properly.

(23) ITEM 78, BODY, LOAD AND TARPAULIN. Inspect load and equipment for damaged, loose, or missing parts. See that ammunition is secure in ammunition compartment, and that tarpaulin is secure and not damaged.

(24) ITEM 81, GUN MOUNTING. Make sure gun mounting is secure and in good condition.

18-TON HIGH SPEED TRACTOR M4

(25) Item 62, *Tighten.

(a) Tighten all bolts, nuts or cap screws observed to be loose.

(b) Weekly. Tighten all vehicle assembly mountings, bolts, nuts or cap screws that are loose.

(26) Item 83, *Lubricate as Needed.

(a) Lubricate all points requiring daily lubrication as shown on Lubrication Guide (figs. 17 and 18). Wipe all dirt from fittings before applying lubricant. Report any missing fittings.

(b) Weekly. Lubricate all points indicated on vehicle lubrication chart as requiring attention on a weekly or hourly basis, or any points that conditions and experience indicate additional lubrication is necessary.

(27) Item 84. *Clean Engine and Vehicle.

(a) Clean dirt and trash from inside cab, body, hull, and ammunition compartment. Remove excess dirt and grease from exterior of vehicle and engine.

(b) Weekly. Wash vehicle and remove all dirt and excessive grease. If washing is impractical, wipe as clean as possible using care not to create bright spots which would cause glare. Clean engine and power train thoroughly.

(28) Item 85, *Tools and Equipment.

(a) Inspect to see that all tools and equipment assigned to vehicle are present and properly stowed or mounted.

(b) Weekly. Clean all tools and equipment of rust, mud, or dirt and see that they are in good condition. Report missing or unserviceable items. CAUTION: *When there is danger of low enough temperature to cause difficulty in starting, the heating stove and canvas cover should be used.* Mount heater below the hull and cover the vehicle completely with appropriate tarpaulins to hold heat around tractor.

Section VII

LUBRICATION

	Paragraph
Introduction	30
Lubrication Guide	31

30. INTRODUCTION.

a. Lubrication is an essential part of preventive maintenance, determining to a great extent the serviceability of parts and assemblies.

31. LUBRICATION GUIDE.

a. General. Lubrication instructions for this materiel are consolidated in a Lubrication Guide (figs. 17 and 18). These specify the points to be lubricated, the frequency of lubrication, and the lubricant to be used. In addition to the items on the Guide, other small moving parts, such as hinges and latches, must be lubricated at frequent intervals.

b. Supplies. In the field it may not be possible to supply a complete assortment of lubricants called for by the Lubrication Guide to meet the recommendations. It will be necessary to make the best use of those available, subject to inspection by the officer concerned, in consultation with responsible ordnance personnel.

c. Lubrication Notes. The following notes apply to the Lubrication Guide (figs. 17 and 18). All note references in the Guide itself are to the step below having the corresponding number.

(1) FITTINGS. Clean before applying lubricant. Lubricate until new lubricant is forced from the bearing, unless otherwise specified. CAUTION: *Lubricate after washing tractor.*

(2) AIR CLEANERS. Every 8 hours or more frequently if required, empty engine air precleaner. Every 8 hours, check level of oil in engine oil bath air cleaner and air compressor oil bath air cleaner (on late models only), and, if necessary, add engine oil to correct level. Every 8 to 32 hours, depending on operating conditions, clean and refill oil reservoir with engine oil. Remove entire air cleaner twice a year, wash and reoil. Clean air pipes and reassemble. Keep all connections tight. Every 100 hours (early models only), remove filter mat in air compressor air filter and wash and saturate with engine oil. Proper maintenance of air cleaners is essential to prolonged engine life.

(3) CRANKCASE. To fill, remove one crankcase breather and filler cap. Drain only when engine is hot. Refill to "FULL" mark on gage. Run engine a few minutes and recheck oil level. CAUTION: *Be sure pressure gage indicates oil is circulating.* See Table. Every 48 hours, remove crankcase breathers, wash and saturate with engine oil.

18-TON HIGH SPEED TRACTOR M4

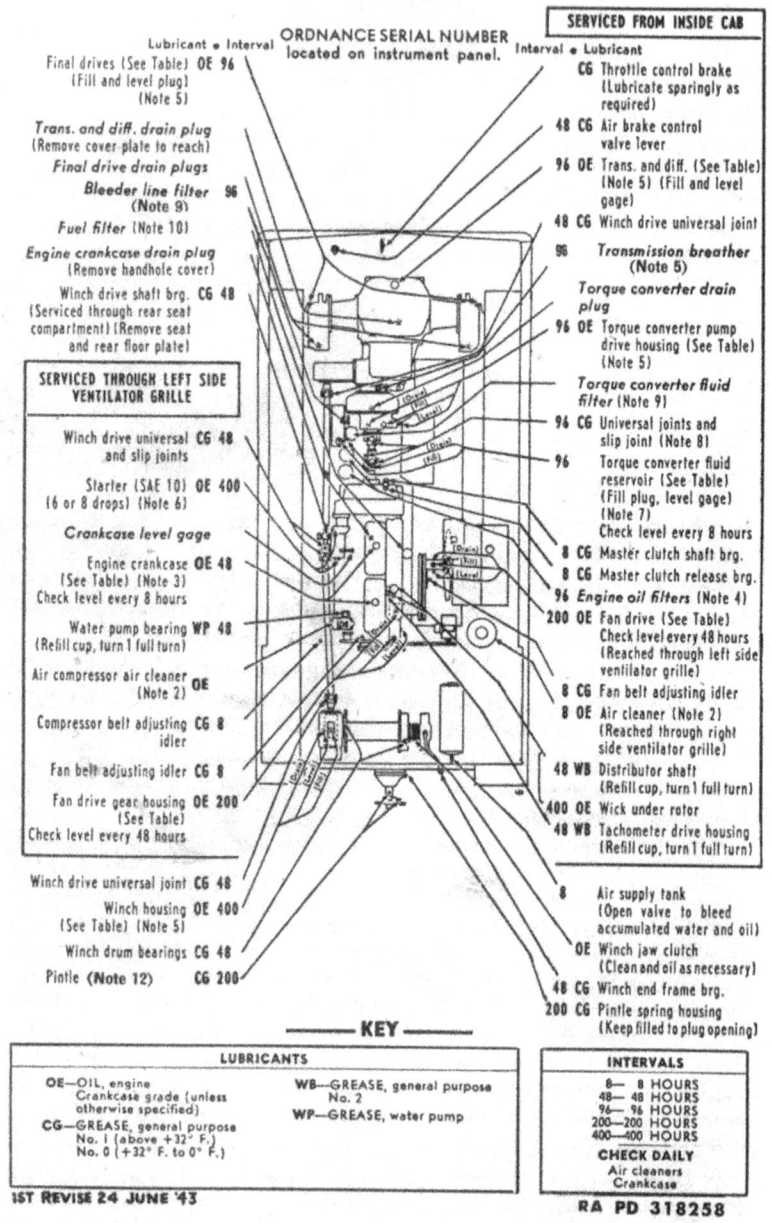

Figure 17—Lubrication Guide—Engine Transmission, Differential, Final Drives and Winch

LUBRICATION

TABLE OF CAPACITIES AND LUBRICANTS TO BE USED

UNIT	CAPACITY (Approx.)	LOWEST EXPECTED AIR TEMPERATURE		
		+32° F. and above	+32° F. to 0° F.	Below 0° F.
Crankcase (without filters)	18 qt.	OE SAE 30	OE SAE 10	Refer to OFSB 6-11
Trans. and Diff.	28 qt.			
Final Drives (each)	10 qt.			
Winch Housing	3 qt.			
Fan Drive Gear Housing	7/8 qt.			
Fan Drive	1/2 qt.			
Torque Converter Pump Drive Housing	1 qt.			
Torque Converter System	34 qt.	Diesel Fuel		OFSB 6-G-150 (122)

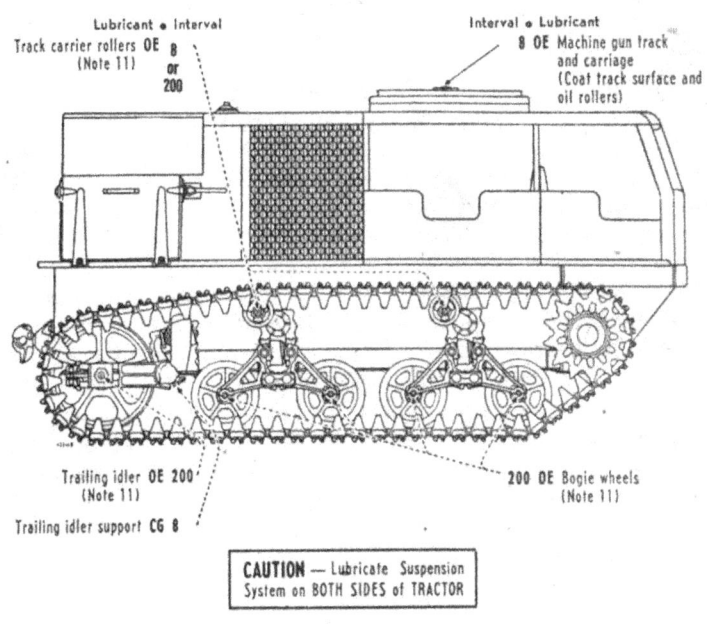

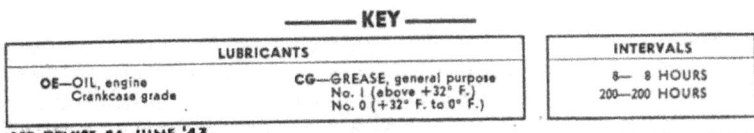

KEY

LUBRICANTS	INTERVALS
OE—OIL, engine Crankcase grade	8— 8 HOURS
CG—GREASE, general purpose No. 1 (above +32° F.) No. 0 (+32° F. to 0° F.)	200—200 HOURS

1ST REVISE 24 JUNE '43

RA PD 318257

Figure 18—Lubrication Guide—Track Suspension

TM 9-785

18-TON HIGH SPEED TRACTOR M4

(4) ENGINE OIL FILTERS. Every 96 hours, or every second oil change, install new element. After installing new element, refill crankcase to "FULL" mark on bayonet gage. Run engine a few minutes and recheck oil level.

(5) GEAR CASES. Every 8 hours, check level with tractor on level ground and, if necessary, add lubricant to correct level. Drain, flush and refill at intervals indicated on Guide. To flush, fill cases to about one-half capacity with engine oil, SAE 10. Operate mechanism within cases slowly for several minutes and redrain. Replace drain plugs and refill cases to correct level with lubricant specified on Guide. After refilling transmission and differential, operate vehicle a short distance, then check oil level and add lubricant to bring level to "FULL" mark. Every 96 hours, remove transmission breather, wash and saturate with used crankcase oil or new crankcase grade engine oil.

(6) CRANKING MOTOR. When the cranking motor is removed, lubricate the outboard bearings with engine oil, SAE 10, and repack the cranking motor reduction gear case with general purpose grease, No. 2.

(7) TORQUE CONVERTER FLUID RESERVOIR. Every 8 hours check fluid level. If necessary, add Diesel fuel to keep level between "LOW" and "FULL" mark on bayonet gage. (Later models will have level petcocks). CAUTION: *Do not fill above "FULL" mark.* Every 8 hours, open drain cock in reservoir to drain accumulated sediment. Every 96 hours, drain and refill torque converter system as explained in paragraph 116 b.

(8) UNIVERSAL JOINTS AND SLIP JOINT. Remove plug in universal joints, insert fitting and lubricate. Replace plug. Slip joint is equipped with a fitting; lubricate until new lubricant is forced from the spline.

(9) TORQUE CONVERTER FLUID FILTERS. Every 96 hours, or more often if filter on fluid reservoir (fig. 126) becomes inoperative, remove filter case and renew element in filter. Install new gasket when element is replaced. Renew element in converter radiator bleeder line filter (fig. 122) every 96 hours, or more often if filter becomes inoperative. Clean body of filter thoroughly and check to make sure orifice in front end of filter body is open before installing new element.

(10) FUEL FILTER. Every 96 hours or more often if filter becomes inoperative, remove and wash filter element. Every 48 hours, open drain plug to remove accumulated sediment and water.

(11) TRAILING IDLERS, BOGIE WHEELS AND TRACK SUPPORT ROLLERS. Clean end of shafts and nozzle of AC flushing lubricator. Remove plugs, insert nozzle as far as it will go into shaft. Lubricate trailing idlers with 14 strokes each, bogie wheels with 6 strokes each, and track support rollers with 6 strokes each. Clean and replace plugs. CAUTION: *Before lubricating, make sure that the lubricator*

LUBRICATION

is delivering the full amount of lubricant. Track support rollers on tractors below serial number M4-301 require lubrication at 8-hour intervals, those on tractors with serial number M4-301 and above require lubrication at 200-hour intervals.

(12) OILCAN POINTS. Every 64 hours, lubricate winch control rod, yoke pins and control lever; steering lever shafts; air and electric brake control linkage; throttle shaft, yoke pins, foot accelerator, carburetor linkage, engine shut-off mechanism, air compressor unloader rocker arm fulcrum pin and pintle (115-mm) with engine oil.

(13) POINTS REQUIRING NO LUBRICATION SERVICE: air compressor, clutch pilot bearing, governor, tracks, bogie pivots.

d. **Points to be Serviced and/or Lubricated by Ordnance Maintenance Personnel.**

(1) GENERATOR. When the generator is removed for maintenance and repairs, remove and wash bearings in dry-cleaning solvent. Allow bearings to dry thoroughly, dip in engine oil, SAE 10, and drain off excess. Repack bearings with general purpose grease, No. 2.

(2) TACHOMETER AND SPEEDOMETER FLEXIBLE DRIVE SHAFTS. Every 400 hours, remove shaft and lubricate with general purpose grease, seasonal grade.

e. **Reports and Records.**

(1) REPORTS. If lubrication instructions are closely followed, proper lubricants used, and satisfactory results are not obtained, make a report to the ordnance officer responsible for the maintenance of the materiel.

(2) RECORDS. Keep a complete record of lubrication servicing.

TM 9-785
32-33

18-TON HIGH SPEED TRACTOR M4

Section VIII

TOOLS AND EQUIPMENT STOWAGE ON TRACTOR

	Paragraph
General	32
Vehicle tools	33
Vehicle equipment	34
Vehicle spare parts	35
Gun tools	36
Gun equipment	37
Gun spare parts	38
Care of equipment	39

32. GENERAL.

a. The items listed in following paragraphs are furnished as standard equipment with each tractor. When tractor is shipped, this equipment is all packed in a box and the box placed on same car as tractor. Boxes and brackets are provided on the tractor to carry this equipment. After tractor is unloaded, the tools and equipment must be removed from the box in which they are shipped and placed in boxes or brackets on tractor before tractor is put into operation.

33. VEHICLE TOOLS.

Tool	Number Carried	Where Carried
a. Pioneer Tools.		
AX (chopping, single bit, 5-lb)	2	One on R front top of cab One on L front top of cab
CROWBAR, 5-ft long, pinch point (41-B-175)	1	Engine cover sheet
HANDLE, mattock (41-H-1286)	2	One on R front top of cab One on L front top of cab
MATTOCK, pick M1 (without handle) (41-M-722)	2	One on R front top of cab One on L front top of cab
SHOVEL, short-handled (41-S-3172)	2	One on R front top of cab One on L front top of cab
SLEDGE, blacksmith dble. face, 10 lb. (41-S-3726)	1	On left rear top of cab
b. Tools, Vehicular.		
BAR, extension, 10-inch (41-B-309)	1	Tool box
BOLT, compressor (volute spring) (41-C-2547-10)	1	Tool box
CHISEL, cold, 1-in. (41-C-1140)	1	Tool box
EXTRACTOR, screw, ½-in.	1	Tool box
FILE, three-square, smooth, 6-in. (41-F-1572)	1	Tool box
FILE, hand, smooth, 8-in. (41-F-1028)	1	Tool box
FIXTURE SET, track connecting (41-F-2997-86)	1	Tool box

TOOLS AND EQUIPMENT STOWAGE ON TRACTOR

Tool	Number Carried	Where Carried
GUN, lubricating, pressure (push type)	1	Tool box
HAMMER, machinist, ball peen 32-oz. (41-H-527)	1	Tool box
HANDLE, combination tee $\frac{1}{2}$-in. sq-dr., 11 inches long (41-H-1504-55)	1	Tool box
HANDLE, wrench, spark plug	1	Tool box
HOLDER, wrench, socket	2	Tool box
JOINT, universal $\frac{1}{2}$-in. sq-dr. (41-J-380)	1	Tool box
LUBRICATOR, flushing	1	Tool box
PLIERS, combination, slip joint 8-in. (41-P-1652)	1	Tool box
PLIERS, side cutting 8-in. (41-P-1977)	1	Tool box
RATCHET, reversible, $\frac{1}{2}$-in. sq-dr. 9-in. (41-H-1505)	1	Tool box
SCREWDRIVER, common 6-in. blade (41-S-1104)	1	Tool box
WRENCH, adjustable single end 8-in. (41-W-486)	1	Tool box
WRENCH, adjustable single end 12-in. (41-W-488)	1	Tool box
WRENCH, engr. dble. hd. alloy stl. $\frac{7}{16}$ x $\frac{1}{2}$-in. (41-W-1000)	1	Tool box
WRENCH, engr. dble. hd. alloy stl. $\frac{9}{16}$ x $\frac{11}{16}$-in. (41-W-1005-5)	1	Tool box
WRENCH, engr. dble. hd. alloy stl. $\frac{19}{32}$ x $\frac{25}{32}$-in. (41-W-1007-20)	1	Tool box
WRENCH, engr. dble. hd. alloy stl. $\frac{5}{8}$ x $\frac{3}{4}$-in. (41-W-1008)	1	Tool box
WRENCH, engr. dble. hd. alloy stl. $\frac{7}{8}$ x $\frac{11}{16}$-in. (41-W-1020)	1	Tool box
WRENCH, engr. dble. hd. alloy stl. 1 x $1\frac{1}{4}$-in. (41-W-1024-1)	1	Tool box
WRENCH, engr. dble. hd. $1\frac{5}{8}$-in. x $1\frac{13}{16}$-in. (41-W-1065-180)	1	Tool box
WRENCH, square plug $\frac{11}{16}$-in.	1	Tool box
WRENCH, hexagon plug $\frac{3}{4}$-in. (41-W-877)	1	Tool box
WRENCH, socket, $\frac{1}{2}$-in. sq-dr. $\frac{7}{16}$-in. hex. (41-W-3005)	1	Tool box
WRENCH, socket, $\frac{1}{2}$-in. sq-dr. $\frac{1}{2}$-in. hex. (41-W-3007)	1	Tool box
WRENCH, socket, $\frac{1}{2}$-in. sq-dr. $\frac{9}{16}$-in. hex. (41-W-3009)	1	Tool box
WRENCH, socket, $\frac{1}{2}$-in. sq-dr. $\frac{5}{8}$-in. hex. (41-W-3013)	1	Tool box
WRENCH, socket, $\frac{1}{2}$-in. sq-dr. $\frac{3}{4}$-in. hex. (41-W-3017)	1	Tool box
WRENCH, socket, $\frac{1}{2}$-in. sq-dr. $\frac{7}{8}$-in. hex. (41-W-3023)	1	Tool box
WRENCH, socket, $\frac{1}{2}$-in. sq-dr. $\frac{15}{16}$-in. hex. (41-W-3025)	1	Tool box
WRENCH, socket, $\frac{1}{2}$-in. sq-dr. 1-in. hex. (41-W-3027)	1	Tool box
WRENCH, socket, $\frac{1}{2}$-in. sq-dr. $1\frac{1}{16}$-in. hex. (41-W-3029)	1	Tool box

TM 9-785
33

TOOLS AND EQUIPMENT STOWAGE ON TRACTOR

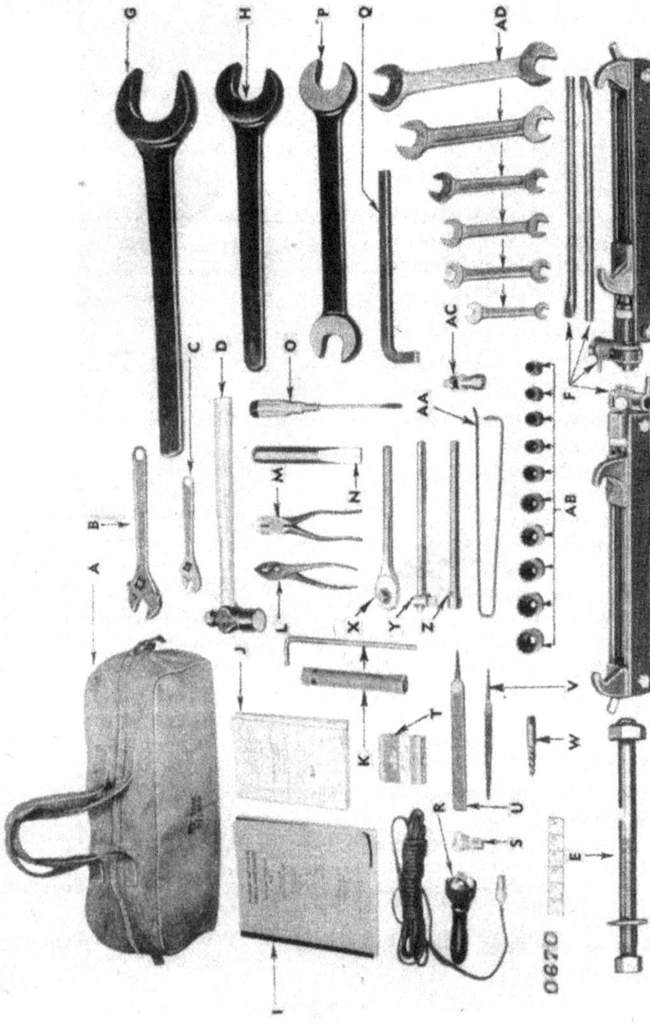

A—TOOL BAG
B—12 IN. ADJUSTABLE WRENCH
C—8 IN. ADJUSTABLE WRENCH
D—BALL PEEN HAMMER
E—VOLUTE SPRING COMPRESSOR BOLT ASSEMBLY
F—TRACK CONNECTING FIXTURE
G—3-3/4 IN. WRENCH
H—2-1/2 IN. WRENCH
I—SERVICE PARTS CATALOG SNL G-150
J—OPERATORS MANUAL TM 9-785
K—SPARK PLUG WRENCH
L—PLIERS
M—SIDE CUTTING PLIERS
N—COLD CHISEL
O—SCREWDRIVER
P—1-5/8 — 1-13/16 IN. WRENCH
Q—SQUARE PLUG WRENCH
R—INSPECTION LAMP
S—BULB
T—FRICTION TAPE
U—FLAT FILE
V—THREE-SIDED FILE
W—SCREW EXTRACTOR
X—RATCHET
Y—SOCKET WRENCH HANDLE
Z—SOCKET WRENCH EXTENSION
AA—SOCKET WRENCH HOLDER
AB—SOCKET WRENCHES 1/2 TO 1-1/8 IN.
AC—UNIVERSAL JOINT
AD—WRENCH SET 7/16 TO 1-1/4 IN.

RA PD 307584

Figure 19—Vehicle Tools

54

TM 9-785
33-35

TOOLS AND EQUIPMENT STOWAGE ON TRACTOR

Tool	Number Carried	Where Carried
WRENCH, socket, ½-in. sq-dr. 1⅛-in. hex. (41-W-3031)	1	Tool box
WRENCH, spark plug, w/o handle	1	Tool box
WRENCH, engr. dble. hd., 1⅝ x 1¹³⁄₁₆-in	1	Tool box
WRENCH, track adjusting 2½-in.	1	Tool box
WRENCH, track adjusting 3¹³⁄₁₆-in.	1	Tool box

34. VEHICLE EQUIPMENT.

Tool	Number Carried	Where Carried
APPARATUS, decontaminating 1½-qt M2	2	On center post in front compartment
BAG, tool	1	Tool box
BLOCK, snatch	1	On top right rear of cab
BUCKET, canvas, folding 18-qt	1	Under seat in rear compartment
CHAIN, tow	1	Around top outside edge of cab
CONTAINER, water 5-gal	3	Inside left engine grille
CONTAINER, 5-gal for engine oil	1	Inside left engine grille
COVER, windshield	2	In tool box
COVER, lamp guard	4	In tool box
EQUIPMENT, cold weather starting kit	1	
EXTINGUISHER, fire, 4 lb CO_2	1	On center post in rear compartment
FLASHLIGHT	1	In tool box
HOSE, tire inflation	1	In tool box
KIT, first aid (24-unit)	1	Recessed in engine wall rear seat compartment
LAMP, bulb, inspection	1	In separate box in tool box
LAMP, inspection	1	In tool box
MANUAL, spare parts, illustrated (for vehicle)	1	In compartment inside roof over cal. .50 amm. box
MANUAL, technical, TM 9-785	1	In compartment inside roof over cal. .50 amm. box
PINTLE	1	In box
ROLL, blanket	11	Inside seat cushions
STOVE, cooking (Coleman 2-burner)	1	Rear stowage compartment
TAPE, friction ¾-in. wide, 30-ft roll	1	Tool box
TUBE, flexible nozzle	2	Tool box
WAR DEPARTMENT LUBRICATION GUIDE	1	Inside top of tool box
WIRE, soft iron 14-ga 10-ft	1	Tool box

35. VEHICLE SPARE PARTS.

Tool	Number Carried	Where Carried
BELT, fan	3	Tool box
BELT, fan, generator and air compressor	5	Tool box
BULB, head lamp, 12-volt 21-32 C.P. double filament	2	In box in tool box
CLAMP STUD (for air brake lines)	2	Tool box
CONNECTOR END	8	Tool box
CONNECTOR, type 1, flared tube ⅜-in.	1	Tool box

55

TM 9-785
38-39

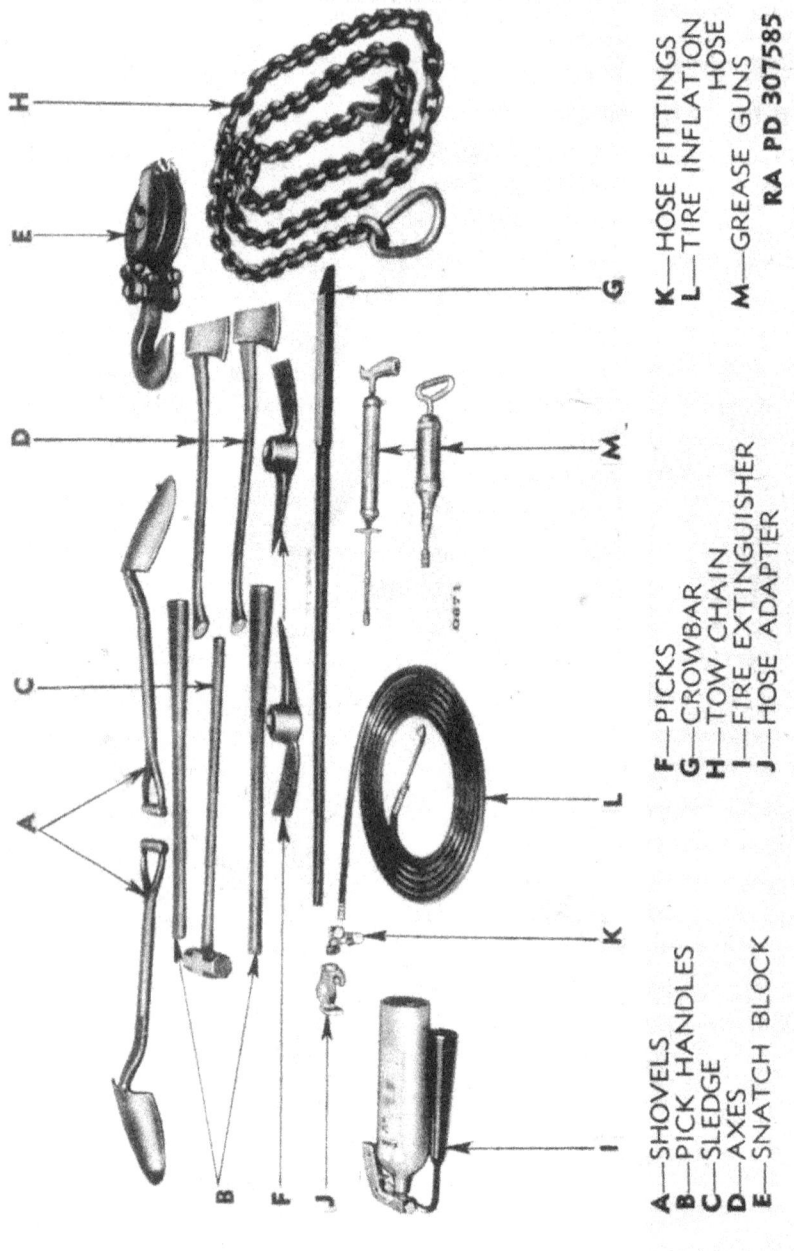

Figure 20—Vehicle Equipment

A—SHOVELS
B—PICK HANDLES
C—SLEDGE
D—AXES
E—SNATCH BLOCK
F—PICKS
G—CROWBAR
H—TOW CHAIN
I—FIRE EXTINGUISHER
J—HOSE ADAPTER
K—HOSE FITTINGS
L—TIRE INFLATION HOSE
M—GREASE GUNS

RA PD 307585

TOOLS AND EQUIPMENT STOWAGE ON TRACTOR

Tool	Number Carried	Where Carried
ELBOW, type 2, flared tube ⅜-in. w/o nuts	1	Tool box
FITTING, push type, str. ⅛-in.—27 NPT male, short	3	Tool box
FITTING, push type, elbow ⅛-in.—27 NPT 90 degrees	3	Tool box
FITTING, push type, elbow ⅛-in.—27 NPT 30 degrees	3	Tool box
GASKET, lubricating plug	2	Tool box
LINK	4	Center front top of cab
NUT, safety, ⅝-in. 18 NF	8	Tool box
NUT, flared tube fitting ⅜-in.	2	Tool box
PINS, shear (for winch)	6	Tool box
PLUG, lubrication	2	Tool box
UNION, flared tube ⅜-in. w/o nuts	1	Tool box
WEDGE	8	Tool box

36. GUN TOOLS.

EXTRACTOR, ruptured cartridge	1	In spare cal. .50 M2 box
OILER, filling, oil buffer	1	
ROD, jointed, cleaning M7	1	
WRENCH, combination M2	1	

37. GUN EQUIPMENT.

BAG, metallic belt link	1	On gun carriage (cal. .50)
BOX, ammunition cal. .50 M2	5	In compartment at driver's left
BOX, ammunition cal. .50 M2 (empty for M.G. spares)	1	In compartment at driver's left
BOX, ammunition cal. .50 (50 rounds)	1	
BRACKET, rifle, universal	1	Inside roof in front of driver
BRUSH, cleaning cal. .50 M4	4	In spare cal. .50 M2 box
CASE, cleaning rod M15	1	
CHUTE, metallic belt link, M1	1	On gun carriage (cal. .50)
COVER, cal. .50 machine gun mount	1	On machine gun
COVER, spare barrel, M13, 45"	1	On barrel
COVER, tripod mount, mach. gun cal. .50 M3	1	On tripod
ENVELOPE, spare parts, M1 (w/o contents)	2	In spare cal. .50 M2 box
MANUAL, field, for cal. .50 M.G. M2	1	In compartment inside roof over cal. .50 amm. box

38. GUN SPARE PARTS.

BARREL, assembly	1	Inside rear personnel comp't.
DISK, buffer	1	In spare cal. .50 amm. box M2
EXTENSION, firing pin assembly	1	In spare cal. .50 amm. box M2

TM 9-785
38-39

18-TON HIGH SPEED TRACTOR M4

Tool	Number Carried	Where Carried
EXTRACTOR, assembly	1	In spare cal. .50 amm. box M2
LEVER, cocking	1	In spare cal. .50 amm. box M2
MOUNT, ring M32 (w/o frame) composed of:	1	
Carriage	1	On ring mount
Cradle	1	On ring mount
Pintle	1	On ring mount
PIN, cotter, belt feed lever pivot stud	1	In spare cal. .50 amm. box M2
PIN, cotter, cover pin	1	In spare cal. .50 amm. box M2
PIN, cotter, switch pivot	2	In spare cal. .50 amm. box M2
PIN, firing	1	In spare cal. .50 amm. box M2
PLUNGER, belt feed lever	1	In spare cal. .50 amm. box M2
ROD, driving spring w/spring assembly	1	In spare cal. .50 amm. box M2
SLIDE, belt feed group, consisting of:		
Arm, belt feed pawl	1	In spare cal. .50 amm. box M2
Pawl, feed belt, assembly	1	In spare cal. .50 amm. box M2
Pin, belt feed pawl, assembly	1	In spare cal. .50 amm. box M2
Slide, belt feed, assembly	1	In spare cal. .50 amm. box M2
Spring, belt feed pawl	1	In spare cal. .50 amm. box M2
SLIDE, sear	1	In spare cal. .50 amm. box M2
SPRING, belt holding pawl	1	In spare cal. .50 amm. box M2
SPRING, cover extractor	1	In spare cal. .50 amm. box M2
SPRING, locking barrel	1	In spare cal. .50 amm. box M2
SPRING, sear	1	In spare cal. .50 amm. box M2
SPRING, belt feed lever plunger	1	In spare cal. .50 amm. box M2
STUD, bolt	1	In spare cal. .50 amm. box M2

39. CARE OF EQUIPMENT.

a. An accurate record of all tools and equipment should be kept in order that their location and condition may be known at all times. Items becoming lost or unserviceable should be immediately replaced. All tools and equipment should be cleaned and in proper condition for further use before being returned to their location.

PART TWO—VEHICLE MAINTENANCE INSTRUCTIONS

Section IX

MAINTENANCE ALLOCATION

	Paragraph
Scope	40
Allocation of maintenance	41

40. SCOPE.

a. The scope of maintenance and repair by the crew and other units of the using arms is determined by the availability of suitable tools, availability of necessary parts, capabilities of the mechanics, time available, and the tactical situation. All of these are variable and no exact system of procedure can be prescribed.

41. ALLOCATION OF MAINTENANCE.

a. Indicated below are the maintenance duties for which tools and parts have been provided for the using arm and ordnance maintenance personnel. Replacements and repairs which are the responsibility of ordnance maintenance personnel may be performed by using arm personnel when circumstances permit, within the discretion of the commander concerned. Echelons and words as used in this list of maintenance allocations are defined as follows:

FIRST AND SECOND ECHELON: Table III AR 850-15	Operating organization driver, operator or crew, companies and detachments, battalions, squadrons, regiments, and separate companies and detachments (first and second echelons, respectively).
THIRD ECHELON: Table III AR 850-15	Technical light or medium maintenance units, including Post and Port Shops.
FOURTH ECHELON: Table III AR 850-15	Technical heavy maintenance and field depot units including designated post and service command shops.
FIFTH ECHELON: Table III AR 850-15	Technical base units.
SERVICE: (Including preventive maintenance) par. 24 a (2) and (3) in part. AR 850-15	Checking and replenishing fuel, oil, grease, water and antifreeze, air, and battery liquid; checking and tightening nuts and bolts; cleaning.

18-TON HIGH SPEED TRACTOR M4

REPLACE:
Par. 24 a (5)
AR 850-15

To remove an unserviceable part, assembly, or subassembly from a vehicle and replace it with a serviceable one.

REPAIR:
Par. 24 a (6) in part
AR 850-15

To restore to a serviceable condition, such parts, assemblies or subassemblies as can be accomplished without completely disassembling the assembly or subassembly, and where heavy riveting, or precision machining, fitting, balancing, or alining is not required.

REBUILD:
Par. 24 a (6)
AR 850-15

Consists of stripping and completely reconditioning and replacing in serviceable condition any vehicle or unserviceable part, subassembly, or assembly of the vehicle, including welding, riveting, machining, fitting, alining, balancing, assembling, and testing.

RECLAMATION:
AR 850-15
Par. 4 (c) in part CIR. 75, dated 16 March '43

Salvage of serviceable or economically repairable units and parts removed from vehicles, and their return to stock. This includes the process which recovers and/or reclaims unusable articles or component parts thereof and places them in a serviceable condition.

NOTES: (1) Operations allocated will normally be performed in the echelon indicated by **X**.

(2) Operations allocated to the third echelon as indicated by E may be performed by these units in emergencies only.

(3) Operations allocated to the fourth echelon by E are normally fifth echelon operations. They will not be performed by the fourth echelon, unless the unit is expressly authorized to do so by the chief of the service concerned.

CLUTCH ASSEMBLY

	2nd	3rd	4th	5th
Clutch assembly, master—service (adjust)	X			
Clutch assembly, master—replace and repair		X		
Clutch assembly, master—rebuild			E	X
Pilot and/or release bearings—replace		X		
Plate, clutch driven—replace		X		
Plate, clutch driven—repair (reline)			X	

TM 9-785
41

MAINTENANCE ALLOCATION

ECHELONS
2nd 3rd 4th 5th

CONVERTER ASSEMBLY, TORQUE

	2nd	3rd	4th	5th
*Converter assembly, torque—replace	*	X		
Converter assembly, torque—repair		X		
Converter assembly, torque—rebuild			E	X
Pump assembly, torque converter—replace	X			
Pump assembly, torque converter—repair		X		
Pump assembly, torque converter—rebuild			E	X

COOLING GROUP

	2nd	3rd	4th	5th
Connections—replace	X			
Fan assembly—service and/or replace	X			
Fan assembly—repair		X		
Fan assembly—rebuild			X	
Radiator assemblies, water and oil—replace	X			
Radiator assemblies, water and oil—repair		X		
Radiator assemblies, water and oil—rebuild			E	X
System, cooling—service	X			
Tanks, surge—replace	X			
Tanks, surge—repair		X		

DRIVE ASSEMBLIES, FINAL

	2nd	3rd	4th	5th
Drive assemblies, final—replace and/or repair	X			
Drive assemblies, final—rebuild			E	X
Sprocket assembly—replace	X			
Sprocket assembly—rebuild			E	X

ELECTRICAL GROUP

	2nd	3rd	4th	5th
Batteries—service, recharge and/or replace	X			
Batteries—repair		X		
Batteries—rebuild			E	X
Cables, battery—replace and/or repair	X			
Coil, ignition—replace	X			
Conduits and junction boxes—replace	X			
Conduits and junction boxes—repair		X		
Lamp assemblies—service and/or replace	X			
Lamp assemblies—repair		X		
Regulator, current and voltage—replace	X			
Regulator, current and voltage—service and/or repair		X		
Regulator, current and voltage—rebuild			X	

---

*The second echelon is authorized to remove and reinstall items marked by an asterisk. However, when it is necessary to replace an item marked by an asterisk with a new or rebuilt part, subassembly or unit assembly, the assembly marked by an asterisk may be removed from the vehicle by the second echelon *only after* authority has been obtained from a *higher* echelon of maintenance.

TM 9-785
41

18-TON HIGH SPEED TRACTOR M4

ECHELONS

	2nd	3rd	4th	5th
ELECTRICAL GROUP (Cont'd)				
Siren assembly—replace	X			
Siren assembly—repair		X		
Siren assembly—rebuild			X	
Switch assemblies—replace	X			
Switch assemblies—repair		X		
Wiper assemblies, windshield—replace	X			
Wiper assemblies, windshield—repair		X		
Wiper assemblies, windshield—rebuild			X	
Wiring—replace	X			
ENGINE				
(Waukesha, 145 GZ)				
Bearings, connecting rod (inserts)—replace		E	X	
Bearings, crankshaft (inserts)—replace		E	X	
Belts—service and/or replace	X			
Block, cylinder—rebuild (recondition)			E	X
Carburetor assemblies—service and/or replace	X			
Carburetor assemblies—repair		X		
Carburetor assemblies—rebuild			X	
Condenser, distributor—replace	X			
Controls and linkage—service and/or replace	X			
Controls and linkage—repair		X		
Cooler, oil—replace	X			
Cooler, oil—repair		X		
Cooler, oil—rebuild			E	X
Crankshaft—rebuild (recondition)			E	X
Distributor assembly—service and/or replace	X			
Distributor assembly—repair		X		
Distributor assembly—rebuild			X	
Drive assembly, fan—replace	X			
Drive assembly, fan—repair		X		
Drive assembly, fan—rebuild			X	
*Engine assembly—replace	*	X		
Engine assembly—repair		X		
Engine assembly—rebuild			E	X
Filter assemblies, oil—replace cartridge	X			
Filter assemblies, oil—replace	X			
Filter assemblies, oil—repair		X		
Flywheel assembly—replace and/or repair		X		

*The second echelon is authorized to remove and reinstall items marked by an asterisk. However, when it is necessary to replace an item marked by an asterisk with a new or rebuilt part, subassembly or unit assembly, the assembly marked by an asterisk may be removed from the vehicle by the second echelon *only after* authority has been obtained from a higher echelon of maintenance.

MAINTENANCE ALLOCATION

ENGINE (Cont'd)

	ECHELONS			
	2nd	3rd	4th	5th
Flywheel assembly—rebuild			E	X
Gaskets, cylinder head, manifold and oil pan—replace	X			
Gear train, timing—replace		X		
Generator assembly—replace	X			
Generator assembly—repair		X		
Generator assembly—rebuild			X	
Governor assembly—adjust and/or replace		X		
Governor assembly—rebuild			E	X
Head, cylinder—replace and/or repair		X		
Head, cylinder—rebuild			E	X
Idler assembly, fan—replace	X			
Idler assembly, fan—repair		X		
Lines and connections, oil (external)—replace and/or repair	X			
Lines and connections, oil (internal)—replace and/or repair		X		
Manifolds—replace	X			
Manifolds—rebuild			X	
Motor assembly, starting—replace	X			
Motor assembly, starting—repair		X		
Motor assembly, starting—rebuild			X	
Pan assembly, oil—service and replace gaskets	X			
Pan assembly, oil—replace and/or repair		X		
Pistons and rings—replace		E	X	
Plugs, spark—service and/or replace	X			
Plugs, spark (two-piece)—repair		X		
Points, breaker, distributor—service and/or replace	X			
Pump assembly, fuel (electric)—replace	X			
Pump assembly, fuel (electric)—repair		X		
Pump assembly, fuel (electric)—rebuild			X	
Pump assembly, oil—replace and/or repair		X		
Pump assembly, oil—rebuild			X	
Pump assembly, water—replace	X			
Pump assembly, water—repair		X		
Pump assembly, water—rebuild			X	
Thermostat—replace	X			
Valves—service	X			
Wiring, ignition—replace	X			

EXHAUST GROUP

| Muffler and exhaust pipes—replace | X | | | |
| Muffler and exhaust pipes—repair | | X | | |

TM 9-785
41

18-TON HIGH SPEED TRACTOR M4

	ECHELONS			
	2nd	3rd	4th	5th

EXTINGUISHER, FIRE

	2nd	3rd	4th	5th
Extinguisher, fire (carbon tetrachloride CCl_4)—service (refill) and/or replace	X			
Extinguisher, fire (carbon tetrachloride CCl_4)—repair		X		
Extinguisher, fire (carbon tetrachloride CCl_4)—rebuild			E	X

FRAME AND BODY GROUP

	2nd	3rd	4th	5th
Boxes, ammunition—replace	X			
Boxes, ammunition—repair		X		
Cab assembly—replace			X	
Cab assembly—repair		X		
Cab assembly—rebuild			E	X
Cushions, seat—replace	X			
Cushions, seat—rebuild			X	
Door assemblies—replace and/or repair		X		
Frame assembly—repair		X		
Frame assembly—rebuild			E	X
Glass—replace		X		
Pintle assemblies—replace	X			
Pintle assemblies—repair		X		
Pintle assemblies—rebuild			X	
Track assembly, gun rail—replace	X			
Track assembly, gun rail—repair		X		
Track assembly, gun rail—rebuild			E	X
Upholstering—replace			X	
Windshield assemblies—replace	X			
Windshield assemblies—repair		X		

INSTRUMENTS

	2nd	3rd	4th	5th
Instruments—replace	X			
Instruments—repair		X		
Instruments—rebuild			E	X

SHAFTS, PROPELLER

	2nd	3rd	4th	5th
Shaft assembly, propeller (w/universal joints)—replace	X			
Shaft assembly, propeller (w/universal joints)—repair		X		
Shaft assembly, propeller (w/universal joints)—rebuild			E	X

SUSPENSION GROUP

	2nd	3rd	4th	5th
Bearings and seals, bogie wheel, idler and track support roller—replace	X			
Bogie components—replace	X			

64

TM 9-785
41

MAINTENANCE ALLOCATION

SUSPENSION GROUP (Cont'd)

	ECHELONS			
	2nd	3rd	4th	5th
Bogie components—repair	X			
Bogie components—rebuild			E	X
Bracket and yoke assemblies, trailing idler—replace	X			
Bracket and yoke assemblies, trailing idler—repair		X		
Bracket and yoke assemblies, trailing idler—rebuild			E	X
Roller assemblies, track supporting—replace	X			
Roller assemblies, track supporting—repair		X		
Roller assemblies, track supporting—rebuild			E	X
Track assemblies—replace and/or repair	X			
Track assemblies—rebuild			E	X
Wheel assemblies, bogie and idler—replace	X			
Wheel assemblies, bogie and idler—repair		X		
Wheel assemblies, bogie and idler—rebuild			E	X

TRANSMISSION AND DIFFERENTIAL ASSEMBLY

	2nd	3rd	4th	5th
Band, brake steering—service and/or replace	X			
Band, brake steering—repair (reline)		X		
Controls and linkage—replace		X		
Controls and linkage—repair		X		
Drum, brake—replace		X		
Pump assembly, transmission oil—replace and/or repair			E	X
Pump assembly, transmission oil—rebuild			E	X
Transmission and differential assembly—replace and/or repair			E	X
Transmission and differential assembly—rebuild			E	X

AUXILIARY EQUIPMENT

BRAKE GROUP, AIR

	2nd	3rd	4th	5th
Compressor assembly, air—service and/or replace	X			
Compressor assembly, air—repair		X		
Compressor assembly, air—rebuild (recondition)			E	X
Controls and linkage—service and/or replace	X			
Controls and linkage—repair		X		
Governor assembly, air pressure—replace	X			
Governor assembly, air pressure—adjust and/or repair		X		
Governor assembly, air pressure—rebuild			E	X
Lines and connections, air—replace	X			
Lines and connections, air—repair		X		
Tank, air—replace	X			
Tank, air—repair		X		
Valve assemblies, air—replace	X			

TM 9-785
41

18-TON HIGH SPEED TRACTOR M4

	ECHELONS			
BRAKE GROUP, AIR (Cont'd)	2nd	3rd	4th	5th
Valve assemblies, air—repair	X			
Valve assemblies, air—rebuild			X	
BRAKE GROUP, ELECTRIC				
Cable assembly, jumper—replace	X			
Cable assembly, jumper—repair		X		
Controller assembly, electric—replace	X			
Controller assembly, electric—repair		X		
Controller assembly, electric—rebuild			X	
Socket assembly, outlet—replace	X			
Socket assembly, outlet—repair		X		
CASE, POWER TAKE-OFF				
Case, power take-off—replace	X			
Case, power take-off—repair		X		
Case, power take-off—rebuild			E	X
Controls and linkage—replace	X			
Controls and linkage—repair		X		
GOOSENECK				
Gooseneck assembly—replace	X			
Gooseneck assembly—repair		X		
Gooseneck assembly—rebuild			E	X
Mount assembly, gooseneck—replace	X			
Mount assembly, gooseneck—repair		X		
Mount assembly, gooseneck—rebuild			E	X
WINCH ASSEMBLY				
Band, safety brake—service and/or replace	X			
Band, safety brake—repair (reline)		X		
Bearings, drive shaft—replace	X			
Cable and hook assembly—replace	X			
Cable and hook assembly—repair		X		
Drum, safety brake—replace	X			
Pin, shear—replace	X			
Shaft assemblies, drive (w/universal joints)—replace	X			
Shaft assemblies, drive (w/universal joints)—repair		X		
Shaft assemblies, drive (w/universal joints)—rebuild			X	
Winch assembly—replace	X			
Winch assembly—repair		X		
Winch assembly—rebuild			E	X

TM 9-785
42

Section X

SECOND ECHELON PREVENTIVE MAINTENANCE

Paragraph
Second echelon preventive maintenance services.......... 42

42. SECOND ECHELON PREVENTIVE MAINTENANCE SERVICES.

a. Regular scheduled maintenance inspections and services are a preventive maintenance function of the using arms and are the responsibility of commanders of operation organizations.

(1) FREQUENCY. The frequencies of the preventive maintenance services outlined herein are considered a minimum requirement for normal operation of vehicles. Under unusual operating conditions such as extreme temperatures, dusty, or sandy terrain, it may be necessary to perform certain maintenance services more frequently.

(2) FIRST ECHELON PARTICIPATION. The drivers should accompany their vehicles and assist the mechanics while periodic second echelon preventive maintenance services are being performed. Ordinarily the driver should present the vehicle for scheduled preventive maintenance service in a reasonably clean condition; that is, it should be dry and not caked with mud or grease to such an extent that inspection and servicing will be seriously hampered. However, the vehicle should not be washed or wiped thoroughly clean, since certain types of defects, such as cracks, leaks, and loose or shifted parts or assemblies are more evident if the surfaces are slightly soiled or dusty.

(3) If instructions other than those contained in the General Procedures in paragraph (4) or the Specific Procedure in paragraph (5) which follow, are required for the correct performance of a preventive maintenance service or for correction of a deficiency, other sections of the vehicle Operator Manual pertaining to the item involved, or a designated individual in authority, should be consulted.

(4) GENERAL PROCEDURES. These general procedures are basic instructions which are to be followed when performing the services on the items listed in the specific procedures. The second echelon personnel must be thoroughly trained in these procedures so that they will perform them automatically.

(a) When new or overhauled subassemblies are installed to correct deficiencies, care should be taken to see that they are clean, correctly installed, properly lubricated and adjusted.

(b) When installing new lubricant retainer seals, a coating of the lubricant should be wiped over the sealing surface of the lip of the seal. When the new seal is a leather seal, it should be soaked in SAE 10 engine oil (warm if practicable) for at least 30 minutes. Then, the leather lip should be worked carefully by hand before installing the seal. The lip must not be scratched or marred.

18-TON HIGH SPEED TRACTOR M4

(c) The general inspection of each item applies also to any supporting member or connection, and usually includes a check to see whether or not the item is in good condition, correctly assembled, secure, or excessively worn. The mechanics must be thoroughly trained in the following explanations of these terms.

1. The inspection for "good condition" is usually an external visual inspection to determine whether or not the unit is damaged beyond safe or serviceable limits. The term good condition is explained further by the following: not bent or twisted, not chafed or burned, not broken or cracked, not bare or frayed, not dented or collapsed, not torn or cut.

2. The inspection of a unit to see that it is "correctly assembled" is usually an external visual inspection to see whether or not it is in its normal assembled position in the vehicle.

3. The inspection of a unit to determine if it is "secure" is usually an external visual examination, a hand-feel, or pry-bar check for looseness. Such an inspection should include any brackets, lock washers, lock nuts, locking wires, or cotter pins used in assembly.

4. "Excessively worn" will be understood to mean worn close to, or beyond, serviceable limits, and likely to result in a failure if not replaced before the next scheduled inspection.

(d) *Special Services.* These are indicated by repeating the item numbers in the columns which show the interval at which the services are to be performed, and show that the parts or assemblies are to receive certain mandatory services. For example, an item number in one or both columns opposite a tighten procedure, means that the actual tightening of the object must be performed. The special services include:

1. Adjust. Make all necessary adjustments in accordance with the pertinent section of the vehicle Operator Manual, special bulletins, or other current directives.

2. Clean. Clean units of the vehicle with dry-cleaning solvent to remove excess lubricant, dirt, and other foreign material. After the parts are cleaned, rinse them in clean dry-cleaning solvent and dry them thoroughly. Take care to keep the parts clean until reassembled, and be certain to keep dry-cleaning solvent away from rubber or other material which it would damage. Clean the protective grease coating from new parts since this material is not a good lubricant.

3. Special lubrication. This applies to lubrication operations that do not appear on the vehicle lubrication chart and to items that do appear on such charts but should be performed in connection with the maintenance operations if parts have to be disassembled for inspection or service.

4. Service. This usually consists of performing special operations, such as replenishing battery water, draining and refilling units with oil, and changing oil filter cartridges.

TM 9-785
42

SECOND ECHELON PREVENTIVE MAINTENANCE

5. *Tighten.* All tightening operations should be performed with sufficient wrench torque (force on the wrench handle) to tighten the unit according to good mechanical practice. Use torque-indicating wrench where specified. Do not over-tighten, as this may strip threads or cause distortion. Tightening will always be understood to include the correct installation of lock washers, lock nuts, and cotter pins provided to secure the tightening.

(e) When conditions make it difficult to perform the complete preventive maintenance precedures at one time, they can sometimes be handled in sections, planning to complete all operations within the week if possible. All available time at halts and in bivouac areas must be utilized if necessary to assure that maintenance operations are complete. When limited by the tactical situation, items with Special Services in the columns, should be given first consideration.

(f) The numbers of the preventive maintenance procedures that follow are identical with those outlined on **W.D. AGO Form No. 462**, which is "The Preventive Maintenance Service Work Sheet for Full Track and Tank-like Wheeled Vehicles." Certain items on the worksheet that do not apply to this vehicle are not included in the procedures in this section. In general, the numerical sequence of items on the work sheet is followed in the Manual procedures, but in some instances there is deviation for conservation of the mechanic's time and effort.

(5) SPECIFIC PROCEDURES. The procedures for performing each item in the 50-hour (500-mile) and 100-hour (1,000 mile) preventive maintenance procedures are described in the following chart. Each page of the chart has two columns at its left edge corresponding to the 100-hour and the 50-hour maintenance respectively. Very often it will be found that a particular procedure does not apply to both scheduled maintenances. In order to determine which procedure to follow, look down the column corresponding to the maintenance due and wherever an item number appears, perform the operations indicated opposite the number.

ROAD TEST

MAINTENANCE		
100-Hour	50-Hour	
		When the tactical situation does not permit a full road test, perform those items which require little or no movement of the vehicle. When a road test is possible, it should be for at least 3, and not over 5, miles.
1	1	**Before-operation Service.** Perform the Before-operation Service outlined in paragraph 27 as a check to determine whether or not the vehicle is in a satisfactory condition to make the road test safely.
2	2	**Instruments and Gages.** Observe all instruments to determine if systems to which they pertain are functioning properly.

69

TM 9-785
42

18-TON HIGH SPEED TRACTOR M4

MAINTENANCE	
100-Hour	50-Hour
3	3
5	5

ENGINE TEMPERATURE GAGE. Normal temperature is 160 to 180°F.

ENGINE OIL PRESSURE GAGE. Normal pressure is 10 to 15 pounds at idling speed and 30 to 40 pounds at normal operating speeds. Lack of pressure or fluctuating pressure requires immediate stopping of engine.

AMMETER. Ammeter should register zero or positive (+) reading during operation. Negative (−) reading may indicate faulty generator, regulator, or other trouble.

FUEL GAGE. Gage should indicate approximate amount of fuel in tank.

SPEEDOMETER AND ODOMETER. Speedometer should indicate vehicle speed without noise or fluctuation. Odometer should record accumulating trip and total mileage.

TACHOMETER. Tachometer should register engine revolutions per minute and count.

AIR PRESSURE GAGE. Gage should indicate 80 to 105 pounds pressure during operation. Low air pressure indicator light should go "ON" when pressure drops below 50 to 60 pounds and should go "OUT" when pressure rises again. Maximum governed pressure is 105 pounds.

TRANSMISSION OIL PRESSURE GAGE. Gage must show at least 5 pounds pressure while tractor is in operation.

TRANSMISSION OIL TEMPERATURE GAGE. Maximum allowable oil temperature is 250°F.

TORQUE CONVERTER FLUID PRESSURE GAGE. Normal operating pressure under load is 40 to 50 pounds.

TORQUE CONVERTER FLUID TEMPERATURE GAGE. If temperature rises above 220°F under normal operating conditions and with normal towed load stop tractor and inspect for cause.

HOUR METER. Meter should be registering during time vehicle is operating. Observe indicator hand for check on operation.

Windshield Wipers and Siren. Inspect siren for secure mounting and proper connections, wiper arms and blades for security and good condition, and wiper motor for smooth operation and full stroke.

Brakes (Steering and Trailer). Before starting vehicle, test operation of steering brakes. Top of lever should move back 6 inches before engagement begins with complete engagement taking place at, or just

70

SECOND ECHELON PREVENTIVE MAINTENANCE

MAINTENANCE	
100-Hour	50-Hour
6	6
7	7
9	9
10	10
11	11

ahead of, vertical position. Both levers should pull back evenly. During operation, test right and left steering brake. On level stretches, observe whether or not vehicle travels straight ahead with both levers released (right or left pull indicates faulty steering brake adjustment or misalinement of track). Test for proper engagement of lever locks. Test action of electric brake, if connected to towed load, by operating electric brake. If towed vehicle has air brakes, test action of air brakes. Test hand operated air brake valve for effective operation. With tractor in motion, disengage clutch when applying air or electric brakes to more readily observe brake action.

Clutch. Clutch pedal should have one and one-half-inches free pedal travel before disengagement of clutch begins. Clutch should not chatter, squeal, or slip. CAUTION: *Do not "ride" clutch.*

Transmission and Torque Converter. Gears must shift smoothly, operate quietly, and not slip out of place during operation. Torque converter should operate quietly.

Engine, Idle. During road test note any tendency of engine to stall when decelerating to shift gears. When vehicle stops, observe whether or not engine will run smoothly at normal idling speed.

ACCELERATION, POWER AND NOISE. Observe if engine has normal acceleration and power in each speed when shifting through entire range. Note unusual engine noise, unusual or excessive "ping" which may indicate early timing or too low octane fuel. Listen for other noises that might indicate damaged, excessively worn, or inadequately lubricated engine parts, or accessories, or loose belts.

GOVERNED SPEED. After engine has warmed up and with vehicle in "LOW" gear, slowly depress accelerator to floor board and by observing tachometer reading see if engine reaches but does not exceed governed speed of 2,100 revolutions per minute under load.

Unusual Noise. Be alert for any unusual noise that may indicate loose parts, damage, or excessive wear in tracks, suspension assembly, or other parts of vehicle.

Temperatures. After covering road test run, observe instrument panel temperature indicators for normal readings. Hand-feel final drives, idlers, bogie wheels, and track support roller hubs. Observe for excessive heating for distance traveled.

18-TON HIGH SPEED TRACTOR M4

MAINTENANCE	
100-Hour	50-Hour
13	13	Leaks. Look within hull and under vehicle for indications of fuel, oil, water, and grease leaks. Trace leaks to source and correct or report.
14	14	Noise and Vibrations. While accelerating and decelerating engine, observe for unusual noises in engine, mountings, accessories, drives, and clutch. Observe if unusual amount of smoke issues from exhaust.
15	15	Track Tension. Observe track tension and adjust if necessary to three-fourths-inch sag between track support rollers.

MAINTENANCE OPERATIONS

16	16	Fuel Pump Test. Attach fuel pump test gage to inlet and outlet fittings of pump. Normal vacuum should be 7 to 14 inches; normal pressure at outlet should be 3½ to 4 pounds.
17	17	Crankcase. Stop engine. Inspect for oil leaks from valve covers, timing gear cover and clutch housing and check oil level. Inspect and tighten crankcase cap screws if necessary. Drain crankcase and refill to "FULL" mark on bayonet gage.
17		SERVE. At 100-hour maintenance service do not start engine again until oil filter service, Item 54, has been completed.
18	18	Side Armor. Examine hull inspection plates to see that all are in good condition and secure. Make sure hull drain plugs are tight.
20	20	Differential, Transmission, Power Take-off, Torque Converter, and Final Drives. Examine these items to see that housings are in good condition and do not leak, that lubricant is up to correct level, and all assembly and mounting bolts are tight.
20		TIGHTEN. Tighten all external assembly and mounting bolts.
21	21	Tracks (Blocks, Connectors, and Wedges). Examine for good condition, correct assembly and secure mountings. Watch for dead blocks, loose or excessively worn blocks or connectors, or bottomed wedges. Replace parts damaged or worn close to, or beyond, serviceable limits. Tighten all track connector wedge nuts securely.
21		On each third 100-hour maintenance service, tracks should be disconnected and removed from sprockets,

SECOND ECHELON PREVENTIVE MAINTENANCE

MAINTENANCE	
100-Hour	50-Hour
22	22
22	
23	23
23	
24	24
24	
25	25
25	
26	26
27	27

track support rollers, and idlers. Inspect related items 22 and 25 at this time, as indicated before reinstalling tracks.

Idlers (Wheels, Arms, Adjusting Mechanism, Lock Nuts and Springs). Examine these items for good condition and correct alinement. Examine idler wheel hubs for oil leaks. Tighten all assembly and mounting bolts securely.

SERVE. At each third 100-hour operation with track removed, examine hub bearings for looseness or end play. Spin idler wheel and listen for unusual noises which might indicate damaged, worn, or inadequately lubricated bearings.

Bogie (Arms, Tie Bars, Springs and Seats, and Brackets). Inspect for general condition and correct assembly. Inspect volute spring for permanent set; two or more coils resting on seat is indication of permanent set and necessitates replacement.

TIGHTEN. Tighten all assembly and mounting bolts.

Wheels (Tires and Rollers). Examine wheels for good condition and secure mountings. Observe for tire adherence to wheels, for cuts, tears or excessive wear. Look for lubricant leaks from wheel or shaft plugs.

SERVE. Test wheels for bearing looseness and end play. Spin wheels, listening for indications of damaged, inadequately lubricated, or worn bearings. At each third 100-hour maintenance service the above operation should be done before tracks are reinstalled. Test support roller bearing for end play and bearing wear. Tighten all assembly and mounting bolts securely.

Sprockets (Hubs, Teeth and Nuts). Examine sprockets for good condition and secure mounting. Inspect sprocket teeth for excessive wear. Excessively worn sprockets should be replaced or reversed. Tighten all mounting bolts securely.

At each third 100-hour operation while track is removed, test for end play of shaft and bearings.

Track Tension. Inspect track tension. Adjust if necessary for three-fourths-inch sag between track support rollers. Tighten bolts and all locking devices.

Top Armor (Body, Paint and Markings, Grilles, Doors, Hull Covers and Latches). Examine to see that these items are in good condition and whether or not the hinges and latches operate properly and are not

18-TON HIGH SPEED TRACTOR M4

MAINTENANCE	
100-Hour	50-Hour
28	28
30	
	31
31	
32	32
33	
34	34
34	

excessively worn. Observe condition of paint for rust, polished surfaces that may cause reflection, and vehicle markings for legibility.

Caps and Gaskets (Fuel and Radiator). Examine for presence, good condition, and snug fit. Open vent in gas cap.

Engine Removal (When Required).
SERVICE. Engine will be removed on the 100-hour maintenance service, only if inspections made in items 6, 9, 13, 14 and a check of organizational records of oil consumption indicate the need.
CLEAN. Clean exterior of engine and dry thoroughly, taking care to keep dry-cleaning solvent away from wiring and electrical equipment. Hot water and soap, not harmful to insulation, should be used when available. Above cleaning and following service in items 31 to 60 will be performed in the best possible manner on engines which do not require removal.

Valve Mechanism. If valve noise indicates loose valves, make adjustment as shown for 100-hour maintenance service.

SERVE. Clean rocker arm covers. Adjust valve stem and rocker arm clearances; intake 0.016 inches, exhaust 0.022 inches (room temperature). Make sure lock nuts are secure after adjustment is completed. Observe rocker arms, shafts, and springs for good condition, secure mounting, and adequate lubrication. Examine rocker arm covers and gaskets for good condition. Wash crankcase breather caps and resaturate with oil. Examine for leaks after starting engine.

Spark Plugs (Gaps and Deposits). Clean and inspect spark plugs for broken insulators. Reset electrodes for 0.025-inch gap (bend only grounded electrodes). Replace faulty plugs.

Compression Test. Test and record on W.D. AGO Form No. 462 compression pressure for each cylinder. Compression of 110 pounds to 130 pounds at normal cranking speed, not varying more than 10 pounds between cylinders, may be considered normal.

Generator and Starter. Examine generator and starter for secure mountings and good connections.

SERVE. Remove inspection covers to see that commutators and brushes are in good condition, not excessively worn, brushes are free in holders, and brush connecting wires are secure and not chafing. Dirty com-

SECOND ECHELON PREVENTIVE MAINTENANCE

MAINTENANCE	
100-Hour	50-Hour
36	36
	36
38	38
39	39
42	42
43	43
44	44

mutators must be cleaned with sandpaper (No. 00). Blow out dust with compressed air.

Distributor (Cap, Rotor, Points, Shaft, and Advance Units). Clean distributor and external attachments and examine for security and serviceability. Inspect for cracks in cap and rotor, loose connection of terminals and cases, and burning of the outer ends of conductor strap. Breaker points must be in good condition, alined, and adjusted to 0.016 to 0.018-inch gap. Replace points and condenser if points are burned or pitted. Test shaft for looseness. Test centrifugal advance to see whether or not camshaft can be rotated by finger force through normal range of movement and whether or not it turns when released without binding or sticking.

SERVE. If breaker plate is excessively worn or dirty, remove distributor, clean in dry-cleaning solvent, dry with compressed air, lubricate, and reinstall in position for timing. When cleaning, remove wick and lubrication cup and clean while removed. Reinstall only after distributor assembly is cleaned and dried. Lubricate removable breaker arm pin, wick, and camshaft with light oil. Fill lubricant cup with cup grease.

Ignition, Wiring and Conduits. Examine wiring for good condition and correct assembly and tighten connections. Observe for chafing against other engine parts. Clean all exposed ignition wiring with dry cloth. Wrap worn spots with friction tape for insulation.

Coil. Examine coil for good condition, cleanliness, and secure mounting.

Breather Caps and Ventilators. Examine transmission and differential, torque converter, fuel tank, and final drive vents for good condition and remove any obstructions from openings. Examine torque converter pump compartment breather, crankcase breathers in rocker arm covers, fan drive housing, and fan shaft housing filler caps for good condition and unobstructed openings.

Air Cleaners. Clean and service engine air precleaner, engine oil bath air cleaner, and compressor air filter (or oil bath air cleaner, as equipped).

Carburetors (Choke, Throttle, Linkage, Governor, Primer). Inspect to see that these items are in good condition, correctly connected, securely mounted, that carburetor does not leak, and that governor is operat-

TM 9-785

18-TON HIGH SPEED TRACTOR M4

MAINTENANCE	
100-Hour	50-Hour
45	45
46	46
48	
49	49
50	50
51	51
51	
52	52
53	53
53	

ing properly (controlling to 2100 revolutions per minute, maximum).

Manifolds. Inspect intake and exhaust manifolds for good condition and security. Examine for leaks due to faulty gaskets or cracks.

Cylinder Head and Gaskets. Examine for indications of oil or water leakage through blow-by around studs, cap screws or gaskets. CAUTION: *Cylinder head should not be tightened unless there is definite indication of looseness or gasket leaks. If tightening is necessary, use torque wrench with 170 to 175 foot-pounds pull on wrench at room temperature.*

Clutch Assembly. On each 100-hour service on those engines that are removed, clean assembly thoroughly and inspect for excessively worn or damaged parts. Lubricate parts before installing on engine.

Water Pump, Fan and Shroud. Examine for good condition and security and see that water pump does not leak. Test fan shaft bearing for end play and blades for proper alinement so they will not interfere with shroud. Test fan drive shafts and bearings for end play and leaks. Adjust drive belts to three-fourths to one-inch deflection. Examine and tighten all mounting bolts.

Accessory Drives (Belts, Pulleys, Shafts and Couplings). Examine items for general condition and secure mounting. Adjust belts for three-fourths to one-inch deflection and tighten all adjustment devices securely.

Engine Compartment. Examine engine compartment in general for cleanliness and secure attachment of all lines, wires, and mountings.

CLEAN. Clean engine compartment as thoroughly as possible. With engine removed, clean out drippings, dirt and refuse; wipe out entire compartment with cloths soaked in dry-cleaning solvent; dry thoroughly.

Engine Oil Cooler, Lines and Fittings. Examine these items for good condition, secure mounting, and indications of leaks.

Fuel Tank, Vents, Lines, and Pump. Examine fuel tank, vents, lines and pump for good condition, secure mountings, and leaks.

TIGHTEN. Tighten fuel tank mountings and fuel line clips or brackets securely. Drain water and sediment

SECOND ECHELON PREVENTIVE MAINTENANCE

MAINTENANCE	
100-Hour	50-Hour

100-Hour	50-Hour	
		from fuel tank by opening drain cock and allowing fuel to drain briefly until it runs clean. Close cock and make sure valve is tight and does not leak. CAUTION: *Use container to catch drainage and avoid spilled fuel. Carefully swab up any spilled fuel.*
54	54	**Engine Oil Filters.** Inspect filters for good condition, security, and leaks.
54		SERVE. Remove filter cartridge, clean case, and install new cartridge and new gaskets, making sure to tighten cover securely. Examine filter for leaks after starting engine.
55	55	**Fuel Filter.** See that fuel filter bowl is in good condition and securely mounted. Drain sediment from bowl.
55		SERVE. Remove bowl and filter element. Wash in dry-cleaning solvent and blow out element with compressed air, being careful not to damage disks. Reinstall, using new gaskets, and examine for leaks after starting engine.
56	56	**Oil Coolers (Transmission, Differential, and Torque Converter Radiator Cores and Lines).** Examine oil and converter fluid cooling radiators and connecting lines for good condition, security and leaks. Clean insects or trash from radiator air passages.
57	57	**Exhaust Pipes and Muffler.** Examine exhaust pipes and muffler for good condition and secure mounting. Observe for indications of exhaust leaks.
57		TIGHTEN. Tighten all mounting bolts and connections securely.
58	58	**Engine Mounting.** Examine for good condition and firm mountings.
58		TIGHTEN. Tighten all accessible mountings and brackets securely.
59		**Clutch Release (Release Yoke, Bearing Assembly).** If engine is removed examine clutch release yoke and release bearing carrier assembly for worn parts. Make sure release mechanism slides freely on release sleeve.
60	60	**Fire Extinguisher.** Examine extinguisher to see whether or not seal is intact. If scale is available, weigh extinguisher for full content. Examine sticker on cylinder for minimum allowable weight. Examine for security of mounting.
61		**Engine (Install, Mounting, Lines and Fittings, Wiring Control Linkage, Oil Supply).** If engine was removed for repair or replacement, reinstall at this

18-TON HIGH SPEED TRACTOR M4

MAINTENANCE	
100-Hour	50-Hour
62	62
63	63
63	
64	64
65	65

time (par. 63). Take care to tighten engine mountings, fuel, oil, water and air line connections and wiring securely. Connect and check control linkage for correct adjustment and lubricate where required. Examine for leaks. Be sure engine crankcase is filled before engine is started.

Radiators (Core, Mounting, Hose, Antifreeze, Record). Examine engine cooling radiator for leaks. Examine radiator mounting, hose, and tubing connections for security. Clean out radiator air passages. Test value of antifreeze with hydrometer and record in space provided on W.D. AGO Form No. 462. When testing antifreeze observe liquid in radiator for presence of rust, oil, or other foreign matter. Flush radiator if necessary. Replenish antifreeze if required.

Battery (Cables, Hold-downs, Carrier, Record of Gravity and Voltage). Clean dirt from top of battery and remove caps. Add clean (drinkable) water to ⅜-inch above separator plates if necessary. CAUTION: *Do not overfill. When terminals are discolored or corroded, clean and grease lightly, then carefully tighten terminals and mounting bolts.* Before adding water to cells, test specific gravity with battery hydrometer and record reading on W.D. AGO Form No. 462. Reading below 1.225 indicates need for recharge. Observe samples of electrolyte for discoloration (reddish-brown) which may indicate battery is being overcharged due to improper generator regulator action. Take voltage reading of each cell and record on W.D. AGO Form No. 462.

SERVE. Make high-rate discharge test of battery according to instructions accompanying test instrument. If differences in readings obtained from cells is more than 30 percent, it should be reported. CAUTION: *A true test is impossible if specific gravity of battery is below 1.225.*

Accelerator. Examine accelerator and linkage for good condition and secure connections. Test throttle for full opening.

Starter (Primer and Instruments). Start engine, observing all precautions outlined in Item 1. Note whether or not starter drive engages and operates properly without excessive noise, if cranking speed is adequate, and if engine starts readily. After engine starts, check instruments for operation, watching oil pressure gage and ammeter for satisfactory indications.

TM 9-785
42

SECOND ECHELON PREVENTIVE MAINTENANCE

MAINTENANCE		
100-Hour	50-Hour	
66	66	**Leaks (Engine Oil, Fuel, and Water).** Look into engine compartment to see whether or not water is leaking from engine cooling system, oil is leaking from engine oil filter or lines, or there are any leaks from fuel system.
67	67	**Ignition Timing.** With engine running, check ignition timing with neon timing light. Observe whether or not automatic control advances spark as engine is accelerated slowly.
68	68	**Regulator Unit (Connections, Voltage, Current and Cut-out).** Observe if unit is in good condition and if mountings are secure. Connect low-voltage circuit tester and test voltage regulator, current regulator, and cut-out for proper generator output control. Follow instructions accompanying test instrument.
69	69	**Engine Idle.** Adjust engine idle speed and mixture.
74	74	**Clutch Pedal.** Test clutch for one and one-half-inch free travel and whether or not pedal return spring brings pedal back to proper released position.
75	75	**Brakes (Steering, Levers, Latches, Linkage).** Examine steering brake levers, ratchets and linkage and see that they are in good condition. Test brake levers for even and proper adjustment (to offer resistance just ahead of vertical position). Inspect trailer air brake valve assemblies for leaks with trailer brakes applied and released. See that dummy couplings are in place on hose couplings if no vehicle is being towed and check hose connections if vehicle is being towed. Drain water (condensation) from air reservoir. CAUTION: *Close pet cocks.* NOTE: *Complete drainage is obtained only with air pressure in tank.* Drain after starting engine if no pressure is left from previous operations. Examine electric brake controller for good connections and mounting. Test operation of vehicle when connected to towed vehicle.
77	77	**Differential, Torque Converter, Breather, and Fillers.** Observe differential breathers and fillers for good condition and clear openings.
77		TIGHTEN. Tighten external assembly, and mounting bolts and screws securely.
81	81	**Propeller Shafts (Joints and Alinement, Seals, and Flanges).** Inspect universal joints for proper alinement

79

18-TON HIGH SPEED TRACTOR M4

MAINTENANCE		
100-Hour	50-Hour	
81		or excessive wear. Slip joints should be free, not excessively worn, and well lubricated. Seals of universal joints and slip joints should not leak excessively. TIGHTEN. Tighten universal joint assembly and companion flange bolts securely.
85	85	**Lamps and Switches.** When tactical situation permits, test lamps and switches to see if all lamps light when switches are in "ON" position, that lights burn properly and go out when switches are turned "OFF." Examine lamps for good condition, secure mounting, broken or dirty lenses and discolored reflectors.
85		ADJUST. Adjust aim of head lamp beam.
86	86	**Wiring (Junction and Terminal Blocks and Boxes, Fuses, and Spares).** Examine all exposed wiring and conduits for good condition, secure support, and tight connection at terminals. Junction terminal blocks and boxes should be in good condition and secure, and necessary serviceable fuses and spares in place. Examine pressure and thermal units connected to indicator instruments for security of electrical connections.
88	88	**Radio Bonding (Filter and Condensers).** Examine to see that connections are clean and secure. NOTE: *Any irregularities, except cleaning and tightening, must be reported to Signal Corps personnel.*
126	126	**Guns.** Examine gun ring and mounting for cleanliness, good condition, firm mounting, and necessary lubrication.
129	129	**Spare Gun Barrels and Parts.** Examine spare gun barrels and parts for condition and proper stowage.
130	130	**Tools (Vehicle, Kit and Pioneer).** Check standard vehicle and Pioneer tools against stowage list for presence, good condition, cleanliness, and proper stowage or secure mountings. Equipment mounted on outside of vehicle should have bright or polished surfaces painted or otherwise treated so they will not cause glare or reflection.
131	131	**Equipment and Winch.** Check equipment against vehicle stowage list. WINCH (CLUTCH, BRAKE DRIVE, SHEAR PIN, CABLE AND GUIDES). Inspect winch for good condition, correct assembly, and security. See that clutch moves freely and engages securely, that the winch worm automatic brake lining is in good condition, and brake correctly adjusted. See that shear pin in winch drive shaft is in good condition. Also see that the oil level in the worm

TM 9-785
42
SECOND ECHELON PREVENTIVE MAINTENANCE

MAINTENANCE	
100-Hour	50-Hour
131	
131	
132	132
134	
135	135
136	136
137	137

gear case is at proper level (even with level plug). If level is high remove level plug and drain oil to obtain proper level. Lubricate the winch clutch, shaft, and operating arm with engine oil. Move the clutch back and forth several times during application of lubricant, to be sure it slides freely.

SERVE. Unwind cable and inspect it for broken or frayed strands and for flat or rusty spots.

CLEAN. Clean entire length of the cable with a cloth saturated with mixture of one part engine oil and four parts kerosene. Dry off excess, and as cable is properly rewound on drum, coat it with a thin film of grease or oil. Drain the worm gear case and refill with new lubricant.

Spare Track Blocks. Examine spare track bolts for presence (4 blocks, 8 wedges, 8 connectors), proper condition, and proper stowage.

Decontaminator. Examine decontaminator for good condition and full charge (test by shaking) and remount securely. NOTE: *Decontaminating solution must be removed every 3 months as it deteriorates.* Check tag on decontaminator for last recharge date.

Fire Extinguisher. Check for presence.

Publications and Form No. 26. The vehicle manual and parts list, Lubrication Guide and Standard Accident Form No. 26 should be present, legible, and properly stowed.

Vehicle Lubrication. Lubricate all points of the vehicle in accordance with Lubrication Guide, instructions in Operator Manual and current lubrication bulletins or directives with the exception of those items which have been lubricated while other services were being performed. On any units where disassembly was necessary for inspection, proper lubrication must be performed unless the vehicle is to be dead-lined for repair of that unit. Observe following instructions: use only clean lubricant and keep all lubricant containers covered; replace missing or damaged lubrication fittings, flexible lines, vents or blocks; on all unfilled bushings or joints the lubricant should be applied until it appears at the openings; on all units provided with lubricant container seals, do not force lubricant beyond seals; do not apply more than specified amount of lubricant to starter, distributor, or water pump (to do so may cause a failure); wipe off excess lubricant that

18-TON HIGH SPEED TRACTOR M4

MAINTENANCE	
100-Hour	50-Hour
138	138
139	139

may drip onto rubber parts, into hull, or may detract from vehicle's appearance.

Modifications (FSMWO'S) Complete. Inspect vehicle to determine that requirements of all Field Service Modification Work Orders have been performed.

Final Road Test. Make final road test, rechecking items 2-15 inclusive. Be sure to recheck lubricant level in transmission, differential, torque converter and final drives. Make sure level and drain plugs are tightly installed and not leaking. Confine road test to minimum distance necessary to make satisfactory observations. Correct, or report, all deficiencies found during final road test.

TM 9-785
43

Section XI

ORGANIZATION TOOLS AND EQUIPMENT

	Paragraph
Organization tools and equipment	43

43. ORGANIZATION TOOLS AND EQUIPMENT.

a. Standard Tools and Equipment. All standard tools and equipment available to second echelon are listed in SNL N-19.

b. Special Tools. The following special tools are available for performing specific maintenance operations.

Tool	Federal Stock Number
Eyebolt, engine-lifting, ⅝-in.—18 NF3	41-B-1586-170
Fixture, track connecting and link pulling, right-hand and left-hand, consisting of:	41-F-2997-86
Fixture, track connecting and link pulling, right-hand	41-F-2997-388
Fixture, track connecting and link pulling, left-hand	41-F-2997-389
Nut, eyebolt, ⅝-in.—18 NF3	41-N-749-600
Puller, slide hammer type, final drive sprocket pinion and trailing idler pivot shaft	41-P-2957-100
Remover and replacer, bogie wheel shaft	41-R-2378-598
Sling, engine-lifting	41-S-3831-810
Tool, unlocking bogie wheel bearing retainer	41-T-3380-30
Wrench, pronged, bogie wheel bearing retainer nut	41-W-3625-130

TM 9-785
44-45

18-TON HIGH SPEED TRACTOR M4
Section XII
TROUBLE SHOOTING

	Paragraph
Introduction	44
Engine	45
Starting system	46
Ignition system	47
Fuel system	48
Cooling system	49
Lubrication system	50
Generator and regulator	51
Power train	52
Tracks and suspensions	53
Winch and power take-off	54
Trailer brake controls	55
Electrical system (lamps, switches, wiring)	56

44. INTRODUCTION.

a. The trouble shooting for the entire vehicle is given in this section.

b. The engine trouble shooting paragraph traces trouble to a "system" affecting engine performance; for example, the fuel, ignition, starting, etc. The symptoms manifested by the engine when in operation are also discussed in the engine trouble shooting paragraph. In order to trace the trouble to one or more defective components of a "system," it is necessary to refer to the pertinent paragraph of this section when the defective system or systems have been located.

c. The power train trouble shooting section includes everything from the engine flywheel to the driving sprockets.

d. The material given in this section applies to operation of the vehicle under normal conditions. If extreme conditions of temperature occur, it is assumed the operator of the vehicle has followed the instructions in section V, "Operation Under Unusual Conditions."

45. ENGINE.

a. Instructions. Part b of this paragraph is the trouble shooting chart for the engine and engine accessory troubles. Part c of this paragraph gives simple engine tests to determine the mechanical condition of the engine. References in part b refer to part c for engine mechanical troubles or to pertinent paragraphs in this section for system troubles, or, when the test indicates a specific unit is faulty, to the pertinent paragraph in this manual. For engine lubrication system troubles, refer to paragraph 50.

TROUBLE SHOOTING

b. **Trouble shooting chart.**

(1) CRANKING MOTOR WILL NOT CRANK ENGINE.

(a) *Lights of Vehicle Stay Bright When Starting Switch Is Closed.*

Possible Cause	Possible Remedy
Open circuit in starting system	Refer to paragraph 46.

(b) *Lights of Vehicle Go Very Dim or Out When Cranking Motor Switch Is Closed.*

Discharged battery	Replace battery (par. 89 j).
Defective starting system	Refer to paragraph 46.

(c) *Cranking Motor Hums When Starting Switch Is Closed.*

Cranking motor drive defective	Replace cranking motor (par. 92 c).
Cranking motor drive sticking due to heavy grease, dirt, etc.	Clean or replace (par. 92 c).
Gear teeth missing from cranking motor gear on flywheel	Notify higher authority.

(2) CRANKING MOTOR CRANKS ENGINE BUT ENGINE DOES NOT START.

(a) *Fuel in Carburetor Does Not Reach Cylinders.*

Empty fuel tank	Fill tank
Fuel valve closed	Open valve
Fuel lines clogged	Examine for clogging
Carburetor not sufficiently choked	Close carburetor choke or, if necessary, adjust (par. 72 c).
Vapor lock	Refer to paragraph 48 b (4).

(b) *No Fuel Reaching Carburetor as Evidenced by No Fuel Flowing from Fuel Line When Disconnected from Carburetor and Engine Is Cranked with Ignition Switch Off.*

Defective fuel system	Refer to paragraph 48 b.

(c) *Excessive Amount of Fuel Being Drawn into Engine Cylinders.*

Engine over-primed or choked	Open choke and throttle. Crank engine for five or six revolutions with "ENG. STOP" pulled out. Wait 30 seconds and repeat operation if engine does not fire.

(d) *Weak or No Spark Emitted from Spark Plug Wire Con-*

18-TON HIGH SPEED TRACTOR M4

nector *When Detached from Spark Plug and Held* $5/16$ *Inch from Engine Block While Engine Is Cranked.*

Possible Cause	Possible Remedy
Ignition switch not closing	Inspect switch.
Defective ignition system	Refer to paragraph 47.

(e) Satisfactory Ignition and Fuel Systems but Engine Backfires or Does Not Fire.

Ignition out of time	Refer to paragraph 47 f.
Inlet or exhaust valve of engine stuck open	Free and lubricate valve or notify higher authority.
Air valve on intake manifold open	Adjust for closing (par. 74 b).

(3) ENGINE IDLES UNEVENLY.

(a) Black Smoke Emitted from Exhaust.

Carburetor idler mixture not properly adjusted	Adjust carburetor idling screw (par. 72 c). If idler mixture fails to adjust, replace carburetor (par. 72 b).

(b) Engine Idles Evenly When Choke Is Partly Closed or a Momentary Even Idle Is Obtained When Engine Is Primed.

Carburetor idle mixture not properly adjusted	Open choke and adjust. If idle mixture adjustment fails to correct trouble, replace carburetor (par. 72 b).
Leaky intake manifold gaskets	Replace gaskets (par. 60).

(c) Engine Idles Unevenly When Fuel System Is Satisfactory and No Leaks Are Present in Inlet Manifold.

Dirty or defective spark plugs	Clean, adjust, or replace defective plugs (par. 65 d).

(d) Engine Idles Unevenly When Fuel and Ignition Systems Are Satisfactory.

Uneven cylinder compression	Engine overhaul indicated.
Leaky cylinder head gasket	Notify higher authority.

(4) LACK OF POWER WHEN ENGINE DOES NOT FIRE ON ALL CYLINDERS.

Spark plugs	Refer to paragraph 65 d.
Engine valves sticking	Refer to c (1) of this paragraph.

(5) LACK OF POWER WHEN ENGINE FIRES ON ALL CYLINDERS.

(a) Engine Overheating.

Defective cooling system	Refer to paragraph 49.
Defective ignition system	Refer to paragraph 47 h.
Carburetor not delivering rated amount of fuel	Refer to paragraph 48 b.
Defective lubricating system	Refer to paragraph 50.

TM 9-785
45

TROUBLE SHOOTING

(b) *Engine Cooling System Satisfactory.*

Possible Cause	Possible Remedy
Defective ignition system	Refer to paragraph 47 h.
Carburetor not delivering rated amount of fuel	Refer to paragraph 48 b.
Air cleaner restricted	Remove and clean (par. 77 c).

(c) *Engine Cooling and Ignition Systems Satisfactory.*

Carburetor not delivering rated amount of fuel	Refer to paragraph 48 b.
Air cleaner restricted	Remove and clean (par. 77 c).

(d) *Engine Cooling, Fuel, and Ignition System Satisfactory.*

Engine has poor compression	Engine overhaul indicated. Perform c (1) of this paragraph.

(6) ABNORMAL NOISE AT IDLE. (refer to c (2), this par.)

(a) *Noise Occurs at Half Crankshaft Speed.*

Loose tappet	Adjust (par. 58).
Broken valve spring	Notify higher authority.
Broken push rod	Notify higher authority.

(b) *Noise Occurs at Crankshaft Speed.*

Loose connecting rod bearing	Notify higher authority.
Loose wrist pin bearing	Notify higher authority.

(7) ABNORMAL NOISE ON HARD PULL.

(a) *Dull Thud Noise at Open Throttle Low Engine Speed.*

Loose main bearings	Notify higher authority.

(b) *Sharp "ping".*

Ignition out of time	Refer to paragraph 47 f.
Carburetor not delivering rated amount of fuel	Refer to paragraph 48 b.
Overheating	Refer to paragraph 49.
Excessive amount of carbon in combustion chamber	Notify higher authority.

c. **Tests to Determine Mechanical Condition of Engine.**

(1) RINGS AND VALVES. If possible, make the following test with the engine warm. Remove a spark plug from each of the engine cylinders. Install a compression gage in number one (rear) cylinder. Open the throttle wide and, with "ENG. STOP" pulled out, crank the engine with starter. Stop cranking the engine when the maximum reading is obtained on the gage. This will take 8 or 10 revolutions of the engine crankshaft. Repeat this process on the remaining cylinders, recording the compression pressure of each cylinder on a piece of paper. If the compression pressure of one or more cylinders is

18-TON HIGH SPEED TRACTOR M4

less than 90 pounds per square inch when the test has been conducted near sea level, the cylinder or cylinders registering this low reading have leaky valves or piston rings. To determine whether or not the rings or valves leak, repeat the above test on the cylinders having poor compression, adding a half teaspoonful of engine oil through the spark plug hole prior to installing the compression gage. If a low reading is obtained, the valves leak. If a satisfactory reading is obtained, the piston rings leak. Sticking valves will sometimes cause leaking valves. In some instances, this condition can be corrected by freeing valves, then lubricate valves while engine is running with a mixture of half light engine oil and half kerosene.

(2) ABNORMAL ENGINE NOISE.

(a) If the noise occurs with each revolution of the engine crankshaft, it is at some point driven by the crankshaft, such as pistons, rings, pins, connecting rods or main bearings, or some member of the engine which is driven at crankshaft speed. A loose main bearing knock is usually a dull thud, more noticeable on a hard pull or quick acceleration.

(b) If the noise occurs once with each two revolutions of the engine crankshaft, the source is at some point driven by the camshaft, such as valves, tappets, etc.

46. STARTING SYSTEM.

a. **Magnetic Cranking Motor Switch Fails to Click When the Cranking Motor Button Switch Is Closed.**

(1) Make sure battery cables and connections are in good condition and tight, also that cables and battery terminals are not corroded.

(2) Test each cell of the battery with a hydrometer. If any cell is less than 1.220 at 80°F, replace the battery (par. 89 j) with a fully charged one and investigate the generating system (par. 51) for the cause and remedy of the discharged battery.

(3) Use a jumper wire to bypass the cranking motor button switch. If the cranking motor solenoid clicks, the wiring between the battery, cranking motor hand switch, and cranking motor solenoid switch is satisfactory. Replace the defective cranking motor switch (par. 92 d). If no click is heard when the cranking motor button switch is bypassed with the jumper wire, use a test lamp to determine whether or not the ignition switch or the circuit between the battery, cranking motor button switch, and magnetic cranking motor switch is defective. Repair or replace defective wires (par. 56 c), and replace the cranking motor button switch (par. 92 c), if defective.

b. **Magnetic Cranking Motor Switch Clicks When the Cranking Motor Button Switch Is Closed.**

TROUBLE SHOOTING

(1) Test the magnetic cranking motor switch with a jumper. If it operates, replace the defective magnetic cranking motor switch (par. 92 d). If it does not operate, proceed with (2).

(2) Remedy all poor connections, short and open circuits in the heavy cable connecting the battery with the cranking motor. Carefully examine and remedy, if necessary, the ground cable circuit of the battery. If the cranking motor does not rotate the engine crankshaft after performing the preceding steps, replace the defective cranking motor (par. 92 c).

47. IGNITION SYSTEM.

a. Remedy all poor connections in ignition wiring system (fig. 45).

b. Remove the distributor cap, dust seal, and rotor. If the movable contact is "frozen", or the contacts are in poor condition, replace them. Adjust the contact point gap (par. 66 c).

c. **Ammeter Registers Slight Reading and Fluctuates When Engine Is Cranked and Ignition Switch Is Closed.**

(1) Perform steps a and b of this paragraph.

(2) Cease cranking the engine and pull "ENG. STOP" out.

(3) Remove the distributor cap from the distributor (par. 66 c). If moisture is present inside the cap, wipe it and the rotor dry. Remove all moisture from the exterior of the cap, the high-tension wires, spark plug porcelains and the high-tension terminal of the coil. Install the distributor cap, hold the ignition switch with hand (leave "ENG. STOP" out) and crank engine with cranking motor, at some time holding end of one spark plug wire about $5/16$-inch away from engine block. If a weak, or no, spark is obtained, proceed with step (4).

(4) Cease cranking the engine and remove the distributor cap from the distributor. Remove the high-tension wire from the center socket of the distributor cap. Rotate the engine crankshaft until the ignition points are closed. Close the ignition switch. Hold the high-tension wire connected to the coil so the free end is about $5/16$-inch from the engine cylinder block and open and close the ignition points with the fingers. If a good spark is obtained, this spark is lost in the distributor cap, rotor, or high-tension wires. Examine the distributor cap and rotor for cracks or a carbon path. If a crack or a carbon path is found, replace the defective distributor cap or rotor (par. 66). If the metal spring of the rotor is broken, or the metal segment is badly burned, replace the rotor. Examine the insulation of the high-tension wiring. If the insulation is broken, oil soaked, or porous, replace the defective high-tension ignition wiring. If a poor spark, or no spark, is obtained, proceed with step (5).

(5) Remove the distributor cap from the distributor. Remove the high-tension wire from the center socket of the distributor cap.

18-TON HIGH SPEED TRACTOR M4

Rotate the engine crankshaft until the points are open. Close the ignition switch. Hold the high-tension wire connected to the coil so the free end is about $5/16$-inch away from the cylinder block. With the bit of a screwdriver, make a ground connection between the movable contact point arm and the ground. If a good spark is obtained from the high-tension wire when the ground connection established by the screwdriver is broken, replace the defective contact points. If a poor spark, or no spark, is obtained, either the condenser, or ignition coil, or both, are defective. Replace each of these units in turn, and if only one is defective, install the serviceable unit which was removed.

d. **Ammeter Registers No Reading and Shows No Discharge When Engine Is Cranked and Ignition Switch Is Closed.** Use a test lamp to find the open circuit in the circuit between the battery and the contact points. First check the ignition switch, and make sure current will pass through it when the switch is closed. Then check the coils and make sure the primary circuit through the coil is continuous. If the circuit through the switch is open when the switch is in the closed position, or if the primary circuit of the ignition coil is not continuous, replace the defective unit. If the circuit is satisfactory to the primary post of the distributor, inspect the grounding cable between the breaker mounting plate and the distributor housing. This cable must be in good order and tightly connected to provide a ground for the mounting plate. Assemble and install the distributor cap. Test the spark delivered to plug. If a poor spark, or no spark, is obtained, perform all of step c of this paragraph.

e. **Ammeter Does Not Fluctuate and Shows a Steady Discharge When Engine Is Cranked and Ignition Switch Is Closed.** Cease cranking the engine, open the ignition switch and remove the distributor cap. Perform step b of this paragraph. Disconnect the condenser lead from the primary post of the distributor. If, when the ignition switch is closed, the ammeter shows no discharge, replace the defective condenser. If the ammeter shows a steady discharge, connect the condenser lead. The trouble has now been traced down to a short circuit in one or more of three items: the wiring, the ignition switch, or the primary circuit in the ignition coil. Use a test lamp to locate the defective unit or units, and correct the short circuit. After correcting the short circuit, assemble the system, and if a good spark is not obtained, observe the ammeter to see whether or not its behavior corresponds to step c or d of this paragraph. Perform the step pertinent to the ammeter's behavior.

f. **Satisfactory Spark Delivered to Spark Plugs but Engine Back-fires or Does Not Fire.**

(1) Check the order of the high-tension wires in the distributor cap and make sure the firing order of the distributor corresponds to the firing order of the engine.

TROUBLE SHOOTING

(2) Check the ignition timing (par. 67).

(3) Test engine compression (par. 45 c (1)).

g. **Engine Idles Unevenly.** Remove the spark plugs; clean and reset the gaps. Replace defective plugs with serviceable ones. Make certain the spark plugs used have the correct heat range (proper type).

h. **Lack of Power When Engine Fires on All Cylinders.**

(1) Check the ignition timing (par. 67).

(2) As a last resort, replace the distributor, since something might have happened to the automatic spark advance to prevent this device from functioning in its intended manner.

48. FUEL SYSTEM.

a. Remove fuel system, disassemble, clean, and install fuel filter (par. 70 b).

b. **No Fuel Reaching Carburetor as Evidenced by Disconnecting Fuel Line from Carburetor and Cranking Engine with Ignition Switch Off.**

(1) Perform step a of this paragraph.

(2) Inspect the fuel pump for fuel leakage through a drain hole in the lower housing. If fuel has been leaking from this hole, the pump has a ruptured diaphragm. Replace the defective pump.

(3) Disconnect both fuel lines from the fuel pump. With the ignition switch open, crank the engine. While the engine is being cranked, close the fitting on the suction line with a finger. If the pump is operating properly, the suction exerted by the pump can be felt by the finger. If the pump does not pass this test, replace it.

(4) The cause of lack of fuel to the carburetor has now been traced to either a plugged fuel line between the fuel tank and the carburetor or vapor lock. The general procedure for removing an obstruction in a fuel line is to disconnect the line at each end, blow it out with compressed air. In the case of vapor lock, the vehicle gradually loses power until the carburetor runs out of fuel. When this happens, if water is available pour it over the fuel pump and the fuel lines subject to engine heat to lower the vapor pressure of the fuel. If no method of cooling these units is available, wait until they cool off.

c. **Lack of Power.**

(1) INSUFFICIENT PRESSURE IN FUEL LINE TO CARBURETOR.

(a) Perform step a of this paragraph.

(b) Connect a liquid pressure gage in the fuel line at the carburetor. Operate the vehicle at near rated engine speed with an open throttle, and observe the fuel pressure registered by the gage. If the

18-TON HIGH SPEED TRACTOR M4

fuel pressure is less than 3½ pounds per square inch, the fuel pump is defective or the fuel line between the tank and carburetor has an obstruction in it. Test the fuel pump as outlined in **a** (3) of this paragraph. If the pump passes this check, disconnect the fuel lines and blow them out with compressed air. If both these operations fail to attain the required fuel pressure at the carburetor, replace the fuel pump (par. 71 b).

(2) REQUIRED PRESSURE IN FUEL LINE TO CARBURETOR. Operate the vehicle near rated speed and with an open throttle. If partly closing the choke produces more power, replace the defective carburetor.

49. COOLING SYSTEM.

a. In the following discussion, it is assumed that the operator has not driven the vehicle far enough after the engine temperature gage has reached the danger zone to boil away sufficient quantity of coolant to interfere with the circulation system.

b. Overheating Not Due to Loss of Coolant.

(1) Remove radiator cap. If cooling system is full and the engine is at normal operating temperature, the water can be observed, or heard, circulating freely. If it is neither seen nor heard, inspect for clogging of lines or radiator, inoperative water pump, or thermostat stuck in closed position (if engine is equipped with thermostat see par. 82 b). Replace water pump if found inoperative, or if radiator is clogged. Clean water lines if clogged. If water is circulating properly check the following possible causes:

(2) Remove any foreign obstruction which would tend to retard the passage of air through the core of the radiator. If insects or dirt have plugged a large number of the air passages through the core, remove these obstructions, working from the inner side of the radiator core, with compressed air or a stream of water.

(3) Check the condition and adjustment of the fan belt. If necessary, adjust or replace it.

(4) Check the ignition timing. If necessary adjust (par. 67).

(5) Continuous operation at open throttle and low engine speed tends to overheat the engine, particularly on a hot day. Keep the engine near rated speed in order to take full advantage of the increased flow of air through the radiator core caused by the fan speed.

(6) Check the carburetor for rated fuel delivery at open throttle as outlined in paragraph 48 c.

c. Overheating Due to Loss of Coolant When Radiator Was Filled Before Starting.

(1) After the engine returns to its normal operating temperature, fill the radiator.

TROUBLE SHOOTING

(2) Inspect and remedy all external leaks of the cooling system, such as hoses, gaskets, etc. If the radiator leaks, replace it with a serviceable unit.

(3) If no external leaks are present, a cracked cylinder head or block, defective cylinder head gasket, defective cylinder sleeve sealing ring, or a ruptured oil cooler element may be the cause. Installation of new parts will be necessary to correct leak if due to above causes.

d. **Loss of Coolant When Vehicle Stands.** If the cooling system loses coolant when the vehicle is idle, and no external leaks are present, inspect oil in crankcase for presence of water due to leaking cylinder head gasket, ruptured oil cooler element, defective cylinder sleeve sealing ring, or cracked cylinder head or block..

50. LUBRICATION SYSTEM.

a. **Detection of Oil Leaks.**

(1) REAR MAIN BEARING. When the vehicle is parked for a few hours after use, spread a large sheet of clean paper directly beneath the oil drain hole in the clutch housing. A few drops of oil appearing upon the paper is a permissible quantity, but not two or three teaspoonfuls. Notify a higher authority if the leak is excessive.

(2) EXTERNAL LINES AND GASKETS. Thoroughly clean all oil line connections and surfaces around engine and oil filter gaskets, and oil drain plugs. After operation of the vehicle, inspect these cleaned areas. Remedy defective connections by installing serviceable ones. Replace defective gaskets. Tighten loose oil drain plugs.

b. **Excessive Oil Consumption.**

(1) Perform step *a* of this paragraph.

(2) Make sure the oil being used corresponds to the recommendation in the Lubrication Guide (fig. 17).

(3) Worn cylinder sleeves, pistons, or rings are main causes of excessive oil consumption. Test for condition of these as outlined in paragraph 45 *c*.

(4) If leak develops in oil cooler element, oil may pass into cooling system. Inspect coolant for presence of oil.

c. **Sudden Loss of Oil Pressure.**

(1) LACK OF OIL. Stop engine immediately and check engine oil level. If the oil level is less than one-third full, fill to full mark and start engine. Stop engine if oil pressure does not register within 30 seconds after starting.

(2) SUFFICIENT OIL. If the oil level is up to the half-full mark or better, notify a higher authority.

TM 9-785
51

18-TON HIGH SPEED TRACTOR M4

51. GENERATOR AND REGULATOR.

   a. Trouble Shooting Chart.

   (1) NO GENERATOR OUTPUT.

Possible Cause	Possible Remedy
Sticking or worn brushes	Notify higher authority or replace generator (par. 90 d).
Gummed or burned commutator	Replace generator (par. 90 d).
Poor contact between commutator and brushes	Refer to b (1).
Commutator rough or out-of-round	Replace generator (par. 90 d).

   (2) UNSTEADY OR LOW OUTPUT.

Possible Cause	Possible Remedy
Loose drive belt	Adjust belt for ¾-in. deflection.
Loose connections	Inspect connections.
Defective wiring	Replace generator (par. 90 d).
Sticking brushes	Correct cause of sticking.
Low brush spring tension	Replace springs or notify higher authority.
Dirty or out-of-round commutator	Clean commutator or replace generator. Refer to b (1) or paragraph 90 d.
High mica on commutator	Replace generator (par. 90 d).

   (3) EXCESSIVE OUTPUT.

Possible Cause	Possible Remedy
Shorted generator field circuit	Replace generator (par. 90 d).
High regulator setting	Notify higher authority or replace regulator.
Defective regulator	Replace regulator.

   (4) NOISY GENERATOR.

Possible Cause	Possible Remedy
Loose pulley	Tighten pulley.
Loose generator mounting bolts	Tighten bolts.
Worn bearings	Replace generator (par. 90 d).
Improperly seated brushes	Notify higher authority or replace generator.

   b. Tests for Generator and Regulator.

   (1) NO GENERATOR OUTPUT. Remove cover band and inspect commutator. If commutator is discolored or dirty, operation may be temporarily restored by holding a strip of No. 00 sandpaper against the commutator with a piece of wood while generator is operating. **DO NOT USE EMERY CLOTH.** If generator still does not operate inspect for sticking or worn brushes. Replace brushes if necessary. Other causes such as rough or out-of-round commutator are not readily apparent and generator will have to be replaced for detailed checking.

TROUBLE SHOOTING

(2) UNSTEADY OR LOW OUTPUT. Unless due to loose drive belt or loose connections, the generator will have to be removed for further checking or rebuilding.

(3) EXCESSIVE OUTPUT. Notify higher authority for checking adjustment or regulator or replace regulator (par. 90 d).

(4) FULLY CHARGED BATTERY AND A HIGH CHARGING RATE. To determine which unit is at fault, with the generator operating at about 2,000 revolutions per minute, disconnect the lead from the regulator "E" terminal. If the output drops off, the regulator is at fault. If the output remains high, the generator is defective and must be replaced.

(5) LOW BATTERY AND LOW OR NO CHARGING RATE. Check by momentarily connecting **ARMATURE** and **FIELD** terminal of regulator with a jumper lead with the generator operating at medium speed. If the output does come up, the regulator should be checked for a low voltage setting or burned or oxidized contacts. Loose connections, defective wiring, or other causes of excessive resistance in the charging circuit may also tend to cause the output to be low, since any one of these would cause the regulator to operate as though the battery were fully charged, even though it were in a discharged condition. If the output remains low with the terminals bridged, then the generator may be considered as being at fault.

52. POWER TRAIN.

a. **Trouble Shooting Chart** (refer to b for procedure to determine location of trouble).

(1) EXCESSIVE NOISE OR VIBRATION IN PROPELLER SHAFT.

Possible Cause	Possible Remedy
Loose bolts in attaching flanges	Tighten or replace bolts.
Worn universal joints	Replace propeller shaft (par. 115 b).

(2) TRACTOR WILL NOT MOVE WITH ENGINE RUNNING AND TRANSMISSION GEARS AND CLUTCH ENGAGED.

Clutch slipping	Adjust or replace clutch (pars. 113 and 114).
Propeller shaft broken	Replace propeller shaft (par. 115 b).
Torque converter inoperative (possibly due to insufficient fluid supply)	Refer to step b. Replace torque converter if necessary (par. 116 c).
Gears or shaft broken or keys sheared in transmission, differential or final drives	Notify higher authority.

(3) EXCESSIVE TEMPERATURE OF CONVERTER FLUID (ABOVE 260°F) REGISTERS ON GAGE.

18-TON HIGH SPEED TRACTOR M4

Possible Cause	Possible Remedy
Tractor not operating in proper speed range	Refer to paragraph 11 d.
Faulty cooling	Inspect for clogged lines, clogged radiator, or inoperative cooling fan.
Operating engine with gearshift in neutral position for extended period	Stop engine..
Air in system	Check fluid supply in reserve tank.
Loss of fluid	Check for loose drain or vent plugs and check for leaks in system with engine running.

(4) ABNORMAL TORQUE CONVERTER FLUID PRESSURE REGISTERS ON GAGE.

(a) Defective Gage.

Check gage by installing tested gage. Replace defective gage (par. 108).

(b) Basic Pressure Too Low.

Insufficient fuel supply	Add fluid to proper level in reserve tank.
Leaks in system	Inspect for leaks.
Filter clogged	Inspect filter, change element if clogged or extremely dirty.
Bypass valve stuck	Inspect valve, replace if inoperative.
Worn or defective auxiliary fluid pump	Replace pump.
Clogged fluid lines	Inspect line from reserve tank to auxiliary pump.
Leaky seals	Check for leakage past seals as explained in step b (7) (b) 6.

(c) Basic Pressure Too High (above 70 Pounds per Square Inch).

Faulty relief valve — Replace valve.

(d) Slow Pressure Response.

Air in system	Bleed air from system.
Bleed line orifice clogged	Clean orifice (see b (7) 1).
Relief valve plunger stuck	Free plunger or replace valve.
Fluid seals leaking	Replace converter (par. 116 c).

(5) EXCESSIVE TEMPERATURE OF TRANSMISSION OIL REGISTERS ON GAGE.

Insufficient or excessive oil supply	Check oil level on bayonet gage.
Lubricating pump for differential inoperative	Replace pump (par. 120).

TROUBLE SHOOTING

Possible Cause	Possible Remedy
Transmission oil lines or cooling radiator clogged or air passages in radiator clogged	Inspect for clogging. Clean or replace lines or radiator (par. 121). Clean trash or dirt from radiator air passages.
Fan belts slipping or broken	Tighten or replace fan belts.
Steering brakes too tight	Adjust brake bands (par. 123).

(6) Low or No Transmission Oil Pressure Registers On Gage.

Possible Cause	Possible Remedy
Insufficient oil supply	Check oil level. Add oil if necessary.
Differential oil pump inoperative	Replace pump (par. 120).
Defective gage	Replace gage (par. 110).

(7) Excessive Noise in Transmission.

Possible Cause	Possible Remedy
Insufficient oil supply	Add oil to proper level.
Sliding gears improperly spaced	Notify higher authority for adjustment of shifter forks on shifter shafts.
Worn or damaged gears or bearings	Notify higher authority.
Overrunning clutch assembly damaged or worn	Notify higher authority.
Idler gear bushings worn	Notify higher authority.

(8) Excessive Noise in Differential.

Possible Cause	Possible Remedy
Differential pinion bushing worn	Notify higher authority.
Brake drums worn between hub thrust and end covers (this condition will be indicated by steering brakes chattering when used to turn tractor)	Notify higher authority.
Gears or bearings worn or damaged	Notify higher authority.

(9) Excessive Noise in Final Drives.

Possible Cause	Possible Remedy
Worn or damaged gears or bearings	Notify higher authority.
Insufficient oil supply in gear case	Add oil to proper level.

(10) Final Drives Running Hot.

Possible Cause	Possible Remedy
Insufficient oil supply in gear case	Add oil to proper level.
Vents in gear cases clogged with dirt	Clean dirt from vents.

TM 9-785
52

18-TON HIGH SPEED TRACTOR M4

(11) TRACTOR WILL MOVE ONLY SLIGHTLY WITH ENGINE AT OPERATING SPEED.

Possible Cause	Possible Remedy
Final drive pinion shaft broken	Notify higher authority.

(12) TRANSMISSION GEARS SLIP OUT OF MESH.

Shifter shaft locking mechanism defective or parts lost	Notify higher authority.
Gears not properly spaced for full mesh or shifter forks worn or bent	Notify higher authority.

(13) TRANSMISSION GEARS LOCKED.

Broken shifter fork or lever broken	Notify higher authority.
Shifter shaft extension loose on shaft	Notify higher authority.

(14) UNABLE TO SHIFT GEARS OR GEARS SHIFT WITH DIFFICULTY.

Shifter fork or lever broken	Notify higher authority.
Shifter shaft locking mechanism defective	Notify higher authority.
Clutch does not disengage fully (drags)	Adjust clutch (par. 113).

(15) FAULTY STEERING.

Steering brakes out of adjustment	Adjust brakes (par. 123).
Brake lining worn out	Replace brake bands (par. 124).
Brake linkage broken or binding	Inspect and lubricate or replace broken parts.

(16) OIL LEAKS.

Broken or damaged oil lines or fittings	Inspect lines and fittings, replace broken lines. Tighten or replace leaking fittings.
Cracked housing	Inspect housings and covers.
Loose covers or plugs or damaged gaskets	Inspect and correct as necessary.

b. **Inspections to Determine Location of Failures or Irregularities in Power Train.**

(1) TRACTOR WILL NOT MOVE WITH ENGINE RUNNING AT OPERATING SPEED WITH TRANSMISSION GEARS AND CLUTCH ENGAGED. Observe if propeller shaft is turning at engine crankshaft speed. If it is not, the clutch is slipping and must be adjusted (par. 113). If propeller shaft is turning at engine speed, inspect torque converter. An insufficient supply of converter fluid will usually be

TROUBLE SHOOTING

indicated by a drop in pressure or no pressure registering on gage. Check fluid level in converter fluid reserve tank. Fill system or add fluid as explained in paragraph 116. With system filled, pressure on gage should be normal. If fluid has been lost, inspect fluid lines and fittings, cooling radiator and converter to determine where leak has occurred. Leaks from radiator or torque converter seals will necessitate replacement of unit. If tractor still fails to move, it is evident that gears or shafts are broken in transmission, differential, or final drives.

(2) TRANSMISSION, DIFFERENTIAL, AND FINAL DRIVE FAILURES. Failure in any one of these assemblies will usually be noted at time of failure. If failure is due to damage to bearings from lack or loss of lubricant, the bearings or shafts will seize and a decided slowing up of tractor and increased load on engine will be noticed. If parts are broken, the broken parts will cause excessive noise and indicate where the trouble can be found. If a shaft breaks in any of these three assemblies, the tractor will fail to move. Failure of the overrunning clutch in transmission will usually cause a grating or squealing noise. Defective shifter shaft locking mechanism failure will be indicated by difficulty in shifting gears or in gears locking. If the operator is unable to shift tractor out of gear or into a desired gear, a broken shifter fork, locked shifter shaft, or shifter shaft extension loose on shaft would probably be found as the cause. Clashing of gears when shifting indicates a dragging clutch. Adjust clutch (par. 113).

(3) STEERING BRAKES. If tractor has a tendency to pull toward one side while traveling (unless on crowned road), it is likely that the brake linkage is binding, or brake band on that side is too tight, or tracks are out of line. Adjust brakes (par. 123), lubricate linkage, and if condition is not corrected, inspect tracks and suspensions for cause. If brakes chatter when engaged, the cause is likely to be due to brake drums being worn, allowing excessive clearance between brake drum thrust and differential end covers and it will be necessary to replace brake drums to correct the condition. Squealing of brakes will indicate worn out linings. If brakes are too tight, high transmission oil temperature is likely to result. Broken linkage, worn lining, or a too loosely adjusted brake would be indicated by failure of the brake, when applied, to turn vehicle.

(4) LACK OF TRANSMISSION OIL PRESSURE. The transmission oil is drawn from a sump on differential housing by a pump on power take-off housing. The oil is pumped through a cooling radiator in radiator assembly and back to the transmission and differential. If no pressure is registered on gage, when tractor is in operation, check level of oil with bayonet gage in differential cover. Add oil to bring level to "FULL" mark if oil level is too low. If oil is at proper level, make sure gage is not defective, then inspect screen in differential

18-TON HIGH SPEED TRACTOR M4

oil sump to make sure it is not clogged. If screen or intake line is not clogged, it is likely the oil pump is inoperative and should be replaced (par. 120).

(5) OIL LEAKS. If it is necessary to add an excessive amount of oil to transmission and differential, inspect for oil leaks. Broken lines or fittings, loose connections, loose bolts, cracked castings, loose covers or damaged gaskets, loose plugs, or leak in oil cooling radiator may be the cause of loss of oil. Replace damaged parts, tighten loose bolts, covers, or connections.

(6) EXCESSIVELY HIGH TRANSMISSION OIL TEMPERATURE. Excessive heating of oil will be caused by insufficient or excessive oil supply, tight or dragging steering brakes, or lack of cooling due to clogged oil lines or cooling radiator or inoperative cooling fan. Inspect for these causes in order named if temperature rises above normal (220°F max.).

(7) IRREGULARITIES IN TORQUE CONVERTER OPERATION.

(a) *Gages.* Because the torque converter is completely enclosed with no moving parts visible from outside, proper performance can be determined only by watching the gages on the instrument panel. Even though the vehicle seems to operate normally, there may be trouble which the gages will reveal. When performance seems to be faulty, the gages, in practically every case short of damage to the converter, will indicate the nature of the trouble.

(b) *Basic Pressure Too Low.* With the engine turning at a speed of 1,000 to 1,200 revolutions per minute, the pressure gage should indicate approximately 35 pounds per square inch minimum. Pressure should respond almost instantly when the engine is started. If pressure fails to reach 35 pounds per square inch, check the following possibilities:

1. Not enough fluid. When temperature of fluid is 100°F, level in reservoir should come to mark on bayonet gage.
2. A serious leak anywhere in the system.
3. Filter badly clogged and bypass valve stuck. Remove body of filter and inspect filter element, and replace if clogged or extremely dirty. Check bypass valve for sticky plunger.
4. Worn or otherwise defective auxiliary fluid pump. Break line between pump and filter, and check flow of fluid with engine operating.
5. Clogged lines. Check line from reserve tank to auxiliary fluid pump for dirt or pinched tubing.
6. Leaky seals. Disconnect front and rear seal drain lines, run engine at 1,000 to 1,200 revolutions per minute and observe fluid leakage past seals. If leakage exceeds about a cupful (½ pt) per minute, it will be necessary to install new seals.

(c) *Basic Pressure Too High.* If at any engine speed, fluid pressure exceeds 70 pounds per square inch on the gage, it indicates

TROUBLE SHOOTING

faulty operation or failure of the pressure relief valve. This valve is installed in the line between the auxiliary pump and the converter inlet, and has a bypass line running to the reserve tank. It is set to maintain a pressure of 35-45 pounds per square inch at 1,000 revolutions per minute or more engine speed, and bypasses excess fluid to the reserve tank when pump discharge pressure exceeds the set valve. Excessive pressures may be built up if the relief valve plunger sticks.

(d) *Temperature Variations.* In normal operation, fluid temperature will be from 150 to 200°F depending upon the loading and upon the temperature of the outside air. Lower temperatures are immaterial, but temperatures in the neighborhood of 260°F and more call for attention, particularly if the vehicle is not being operated in an excessively hot climate. It is safe to operate with temperatures as high as 300°F for short periods of time or in emergencies, but beyond this point it is dangerous. It is necessary to stay well below the flash point of the fluid to avoid the possibility of fire or explosion, and even if this is done, the high temperature causes excessive evaporation and possible chemical reactions which may tend to clog the system or otherwise interfere with proper operation. Excessive temperatures may be due to:

1. *Long continued operation under heavy pull.* Under extreme conditions of load (especially complete stall under full throttle), the converter is forced to slip, which means the entire power input is being converted into heat.

2. *Faulty cooling.* Make sure that air is circulating through the fluid radiator. If the radiator seems reasonably cool despite high temperature on gage, it may mean that cooling lines are clogged. Circulation through the cooler is accomplished by the same fluid velocity which drives the vehicle; so if the converter is functioning, you can be sure the necessary circulating pressure is available.

3. *Continued operation at no load.* If the engine and converter are operated at relatively high speed with the transmission behind the converter in neutral, excessive temperature may prevail as the power absorbed from the engine is transformed into heat by the converter.

4. *Air in system.* Air in the fluid system causes excessive slip which in turn causes undue heating. Check fluid level in reserve tank.

(e) *Loss of Fluid.* Unless there are leaks, the only loss of fluid is the result of evaporation. Hence any appreciable drop in fluid level over a short period of time is a sure indication of a leak. The first things to check are the drain plug and the converter vent plug. If these are tight, the entire system must be checked for leaks. To find leaks, it is necessary to have the system under pressure, which can be accomplished either by running engine and converter at idling speed or by connecting an air line to the system and thus imposing static pressure.

18-TON HIGH SPEED TRACTOR M4

(f) Slow Pressure Response. Failure of pressure to respond almost instantly when engine is started, or sluggish, slow return of pressure to zero when engine is stopped, indicates air in the fluid system. Normal operation at 500 revolutions per minute or more engine speed should vent the air in three to five minutes. If air is still present after such operation, the orifice, at the connection of radiator bleed line to reserve tank, probably requires cleaning.

1. Slow pressure response may also be due to the fact that the relief valve plunger is stuck in such a manner that all of the fluid is bypassing instead of being pumped into the converter, or to excessive fluid leakage at the fluid seals. Break seal drainage lines, run engine and check leakage.

53. TRACKS AND SUSPENSIONS.

a. **Trouble Shooting Chart.**

(1) TRACK OUT OF LINE (WEARING ON SIDES OF WHEELS AND TRACK CONNECTORS).

Possible Cause	Possible Remedy
Trailing idler out of alinement	Aline idler (par. 133 e).
Rubber bushings in track block worn or cut out	Replace defective track blocks.
Track block pins worn more on one end than other	Replace track (par. 133).
Drive sprocket worn unevenly	Replace sprockets (par. 134).

(2) TRACK CONNECTORS LOOSE.

Possible Cause	Possible Remedy
Wedge blocks loose	Tighten wedge bolts.
Wedge bottomed	Replace wedges or blocks.

(3) ONE SIDE OF TRACTOR SAGS.

Possible Cause	Possible Remedy
Volute springs weak, "set", or broken in bogie	Replace spring (par. 136 d).

(4) BOGIE TIRES WEARING UNEVENLY.

Possible Cause	Possible Remedy
Bogie assembly out of line	Inspect for sprung or broken bogie arm or support bracket. Replace bogie (par. 136) if necessary.

(5) OIL LEAKS FROM BOGIE WHEELS, TRAILING IDLERS, OR TRACK SUPPORT ROLLERS.

Possible Cause	Possible Remedy
Oil seals damaged	Replace leaking assembly.
Shaft plugs loose or plug gaskets damaged	Fill wheel with oil, replace plug gaskets and tighten plugs.

b. **Inspections for Items in Trouble Shooting Chart.**

(1) TRACKS. "Dead" blocks can be identified by the fact that one side will rise up above the level of the same side of the other

TROUBLE SHOOTING

blocks when on the upper side of the track, because the rubber bushings have separated from the pins. If track connectors are rubbing on one side of idler, bogie wheel, and track support rollers, or track tends to "climb" wheels or sprockets, the track is out of alinement and trailing idler must be alined (par. 133 c) to correct trouble. Wedge blocks in track connectors will "bottom" after repeated tightening and, in that event, must be replaced.

(2) BOGIE. Oil leaking from wheel indicates damaged oil seals, loose shaft plug, or damaged plug gasket. If oil seals are leaking, replace wheel (par. 136). Uneven wear on tires indicates bent arms, pivot shafts, or support brackets. If tractor sags on one side, inspect volute springs. If two or more coils are resting on spring seat, replace spring (par. 136 d).

(3) TRACK SUPPORT ROLLERS. Oil leaking from rollers indicates leaking oil seals, loose shaft plug, or damaged plug gasket.

(4) TRAILING IDLERS. Track connectors rubbing on side of wheel indicates misalinement of wheel with tractor, or bent or broken idler arm or bracket. Oil leaking from wheel indicates damaged or scored oil seals, loose shaft plugs, or damaged plug gaskets.

54. **WINCH AND POWER TAKE-OFF.**

a. **Trouble Shooting Chart (Winch).**

(1) OIL LEAKAGE.

Possible Cause	Possible Remedy
Ball bearings on worm shaft worn	Replace winch (par. 141).
Cracked case	Replace winch (par. 141).
Oil seal in worm shaft bearing cap worn	Notify higher authority or replace winch (par 141).
Oil seal in brake case worn	Notify higher authority or replace winch (par. 141).

(2) WINCH FAILS TO HOLD LOAD.

Possible Cause	Possible Remedy
Automatic brake spring too loose	Adjust brake spring (par. 140 a).
Brake lining worn or defective	Inspect lining. Replace band if necessary.
Oil on brake lining	Clean oil from linings and determine source of oil.

(3) WINCH BRAKE OVERHEATS.

Possible Cause	Possible Remedy
Automatic brake out of adjustment	Adjust automatic brake (par. 140 a).

(4) WINCH DRUM WOBBLES.

Possible Cause	Possible Remedy
Drum shaft bent	Replace winch (par. 141).

18-TON HIGH SPEED TRACTOR M4

(5) END FRAME AND GEAR CASE WOBBLE.

Possible Cause	Possible Remedy
Large pivot bolts holding each unit to winch base may be loose	Tighten bolts.
Drum shaft may be bent	Replace winch (par. 141).

(6) SLIDING CLUTCH DOES NOT ENGAGE.

Possible Cause	Possible Remedy
Clutch sticks	Lubricate clutch and shaft and inspect linkage.
Clutch jumps out of engagement	Adjust linkage on clutch control rod (par. 140 b).

b. Trouble Shooting Chart (Power Take-off).

(1) NOISE IN GEAR BOX.

Possible Cause	Possible Remedy
Gears worn	Replace power take-off (par. 143).
Bearings loose	Notify higher authority.
Mounting cap screws loose	Tighten cap screws.

(2) SLIPS OUT OF GEAR.

Possible Cause	Possible Remedy
Incorrect adjustment on shifter rod	Adjust rod to obtain full gear mesh.
Worn shifter shaft at locking point	Notify higher authority.
Lock spring and ball worn	Notify higher authority.
Pinion shaft bearings out of adjustment	Notify higher authority.
Gears badly worn	Replace assembly (par. 143).
Pinion shaft splines worn	Replace assembly (par. 143).
Shift fork worn	Notify higher authority.

(3) OIL LEAKS AROUND POWER TAKE-OFF SHAFT.

Possible Cause	Possible Remedy
Outer seal defective	Replace assembly (par. 143).
Bearings out of adjustment	Replace assembly (par. 143).

(4) EXCESSIVE WEAR ON GEARS.

Possible Cause	Possible Remedy
Insufficient lubrication	Check differential oil level.
Bearings out of adjustment	Notify higher authority or replace assembly (par. 143).
Gears not in full mesh	Adjust shifter fork on shifter shaft.

55. TRAILER BRAKE CONTROLS.

a. Trouble Shooting Chart (Air Brake Controls).

(1) SLOW PRESSURE BUILD-UP IN RESERVOIR.

Possible Cause	Possible Remedy
Brake valve leaking	Replace valve assembly (par. 128).
Leaking compressor discharge valves	Notify higher authority.

TM 9-785
55

TROUBLE SHOOTING

Possible Cause	Possible Remedy
Leaking lines or connections	Tighten fittings or replace tubing and fittings.
No clearance at compressor unloader valves	Adjust clearance to 0.010 to 0.015 in. clearance (par. 126 b (3)).
Clogged compressor air filter	Clean filter (par. 126 b (2)). If equipped with oil bath cleaner, service cleaner same as oil bath cleaner for engine.
Worn compressor piston rings	Replace compressor assembly (par. 126 c).
Carbon in discharge line	Replace or clean line.

(2) QUICK LOSS OF RESERVOIR PRESSURE WHEN ENGINE IS STOPPED.

Worn or leaking compressor discharge valves	Replace compressor (par. 126 c).
Tubing or connections leaking	Replace tubing or tighten fittings.
Valve leaking	Replace valve (par. 128).
Air pressure governor leaking	Replace governor.

(3) PRESSURE IN AIR BRAKE SYSTEM RISES ABOVE 105 POUNDS (COMPRESSOR NOT UNLOADING).

Broken unloader diaphragm in compressor cylinder head	Replace compressor (par. 126 c).
Excessive clearance at unloader valves	Adjust for 0.010 to 0.015 in. clearance (par. 126 b (3)).
Excessive carbon condition in compressor cylinder head unloading cavity	Replace compressor (par. 126 c).
Restriction in line from governor to compressor unloading mechanism	Replace or clean tubing.
Air supply valve handle in wrong position	Turn handle to correct position (parallel to body of valve).
Air pressure governor inoperative	Replace governor.

(4) SLOW BRAKE APPLICATION.

Low brake valve delivery pressure	Replace valve assembly (par. 128).
Restriction in tubing or hose	Clean or replace.
Leaking brake valve diaphragm or piston packing cup	Replace valve assembly (par 128).

(5) SLOW BRAKE RELEASE.

Restriction in tubing or hose	Clean or replace.
Faulty brake valve	Replace (par. 128).

TM 9-785
55

18-TON HIGH SPEED TRACTOR M4

(6) INSUFFICIENT BRAKES.

Possible Cause	Possible Remedy
Restriction in tubing or hose	Clean or replace.
Brake lining worn excessively or brakes on towed vehicle out of adjustment	Check hose lines from tractor to towed vehicle. Be sure they are properly connected and that the cut-out cocks in the lines are open (handles crosswise of pipe).
Compressor fails to maintain adequate supply of air pressure	Correct leakage or replace compressor (par. 126 e).

b. **Test for Air Brake System.** In some cases, it will be necessary to have the towed vehicle and hose from it connected to tractor in order to locate causes of improper functioning.

(1) AIR PRESSURE TESTS.

(a) With the motor running, observe at what pressure registered by the air gage the governor cuts out and compression is stopped. This pressure should be approximately 100 to 105 pounds.

(b) Observe at what pressure the governor cuts in and compression is resumed while slowly reducing the air pressure in the reservoir by applying and releasing one of the brake valves. This pressure should be approximately 80 to 85 pounds.

(2) LEAKAGE TESTS.

(a) With the brake system fully charged, the motor stopped and both brake valves in released position, observe the drop in reservoir air pressure registered by the air gage. The drop should not exceed 2 pounds per minute.

(b) With the motor stopped and both brake valves in applied position, observe the drop in reservoir air pressure registered by the air gage. The drop should not exceed 3 pounds per minute.

(c) Check for leaks in lines or connections with soapy water and a clean paint brush.

(3) VALVE DELIVERY PRESSURE TEST. Connect an accurate air test gage to the service line outlet at the rear of the tractor and open the service line cut-out cock. When the foot operated brake valve is depressed to its fully applied position, the air test gage should register approximately full reservoir pressure as registered on the dash gage. When the hand operated brake valve is moved to fully applied position, the air test gage should register at least 60 pounds pressure.

c. **Electric Brake Control System.** Since there are no working parts to the electric brake control units other than the operating linkage for the brake controller, inspection for inoperative units will consist merely of testing for delivery of current from unit to

TROUBLE SHOOTING

unit. If brakes fail to operate, due to faulty control units, proceed with following steps to determine which is inoperative. It will be necessary in most cases to have towed vehicle connected to test for operation of brakes.

(1) Inspect wires from ammeter to resistor, resistor to load control, load control to brake controller, and from brake controller to coupling socket for loose connections and broken or grounded wires.

(2) Clean contacts on load control unit.

(3) If further checking is necessary, check resistor, load control, and brake controller by replacing with new or rebuilt units in order named until inoperative unit is eliminated.

56. ELECTRICAL SYSTEM (LIGHTS, SWITCHES, WIRING).

a. The ignition system, starting system, and generating system are covered in preceding sections. The remaining electrical units of the tractor are: lamps, switches, and wiring.

b. The main light switch (fig. 84) controls the service lights, blackout driving light, and blackout marker lights. In the event any of these lamps fail to burn when switch is turned on, install a new bulb. If the new bulb does not light when switch is turned on, inspect wire leading from ammeter to switch and from the switch to the light and dimmer switch for loose or poor connections, or broken wire. If wires and connections are found to be in good condition, the trouble will be a defective dimmer switch or light switch. If lights go off and on, intermittently, inspect for loose lamp mounting bolts or grounded wire due to frayed insulation. The intermittent lighting is caused by a thermal unit on switch that operates to cut off the current when a short circuit occurs and prevents burning out bulbs. There are no fuses in light wiring system.

c. Refer to paragraph 97 and wiring diagram (figs. 94 and 95) when replacing wires.

d. If lights are dim, inspect for dirty lenses, discolored reflectors, partial shorts in wires, loose lamp mounting bolts, or dirty wire connections or terminals. A weak battery can also cause dim lights.

TM 9-785
57

18-TON HIGH SPEED TRACTOR M4

Section XIII

ENGINE

	Paragraph
Description and tabulated data	57
Valve adjustment	58
Governor	59
Intake and exhaust manifolds	60
Cylinder head gasket replacement	61
Removal of engine	62
Installation of engine	63

57. DESCRIPTION AND TABULATED DATA.

   a. **General Description.** The engine used in this tractor is a 6-cylinder gasoline engine. It is a high compression, 4-cycle, valve-

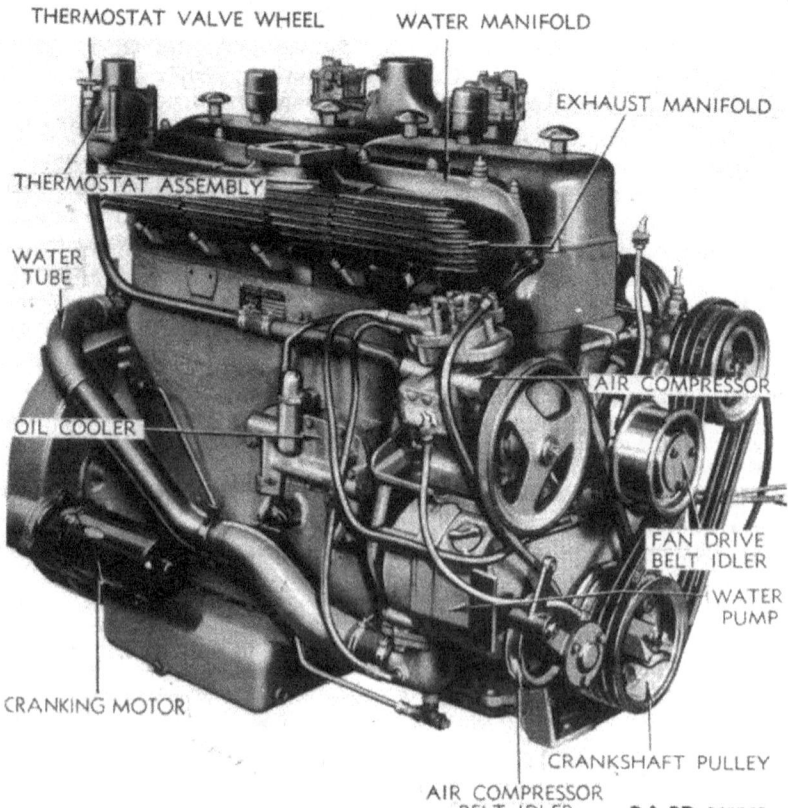

Figure 21—Left Rear View of Engine

108

TM 9-785
57

ENGINE

in-head, water-cooled engine with dual carburetion and electrical ignition system. The engine is mounted in the center of the tractor on rigid supports with flywheel end towards front of tractor. NOTE: *Throughout this manual, unless otherwise noted, the flywheel end of engine will be referred to as the front end of engine.*

b. **Lubrication.** All moving parts of the engine are lubricated by a positive pressure system. There are few outside oil lines, the oil

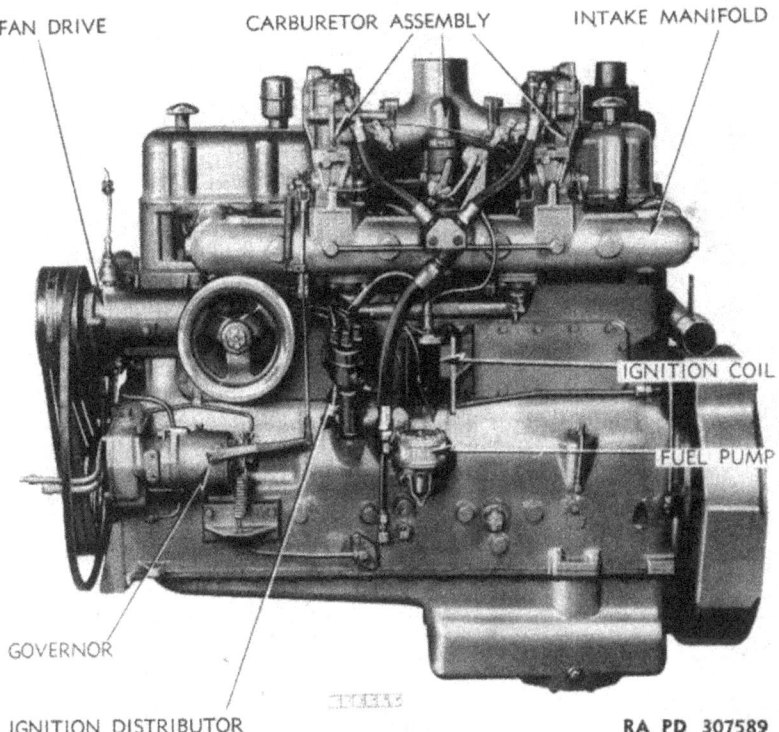

Figure 22—Right Side of Engine

being delivered by a gear type pump to the various operating parts through drilled passages in the cylinder block and head, crankshaft, connecting rods, camshaft, and rocker arm assemblies. Oil pressure is regulated by pressure relief valves located in the oil pump, at oil cooler inlet, and cylinder block oil gallery. Lubricating fittings are provided on some of the attached accessories.

c. **Engine Accessories.** The electric cranking motor, oil cooler, water pump, air compressor, and exhaust manifold are mounted on

18-TON HIGH SPEED TRACTOR M4

left side of engine. The coil, distributor, carburetors, intake manifold, fuel pump, and governor are mounted on right side. An L-shaped fan drive gear assembly is also bolted to right side of engine. The fan drive gear assembly and air compressor are driven by V-belts from a pulley on rear end of engine crankshaft, the generator (mounted on fuel tank) and fan are driven by belts from the pulley on fan drive gear shafts.

d. Tabulated data.

(1) GENERAL.

```
Make and model .............................. Waukesha 145 GZ
Type ....................................................Gasoline
Number of cylinders ........................................ 6
Number of cycles.......................................... 4
Firing order (from rear of engine) ................. 1-5-3-6-2-4
No. 1 cylinder location ................ At end opposite flywheel
Minimum horsepower at 2100 rpm ....................... 210
Bore and stroke ................................... 5 3/8 x 6 in.
Piston displacement ............................... 817 cu in.
Compression ratio .................................... 6 to 1
Maximum torque ..................... 585 ft lbs at 1,500 rpm
Maximum permissible speed (governed)............ 2,100 rpm
Weight (with accessories)................ 2,150 lb (approx.)
Weight (less accessories) ............... 1,800 lb (approx.)
Over-all length ..................................... 55 1/4 in.
Over-all height ....................................... 48 in.
Over-all width ........................................ 35 in.
Crankcase capacity (less lubricating oil filter absorption)...... 5 gal
```

(2) DIRECTION OF ROTATION OF ACCESSORIES (VIEWED FROM FLYWHEEL END).

```
Crankshaft ................................... Counterclockwise
Starter ............................................ Clockwise
Generator .................................... Counterclockwise
Water pump .................................. Counterclockwise
Oil pump (looking down on shaft) ............ Counterclockwise
Distributor (looking down on shaft) ................. Clockwise
Governor ........................................... Clockwise
Air compressor ............................... Counterclockwise
```

(3) RATIO OF ACCESSORY DRIVE TO CRANKSHAFT SPEED.

```
Starter ......................................... 20.65 to 1
Generator ........................................ 1.2 to 1
Water pump ...................................... 0.5 to 1
Oil pump ....................................... 0.77 to 1
Distributor ...................................... 0.5 to 1
Governor ........................................ 1.5 to 1
Air compressor ................................. 0.85 to 1
```

TM 9-785
58

ENGINE

58. VALVE ADJUSTMENT.

a. Correct valve lash is important to insure correct and efcient operation of the engine. Too much valve lash causes excess wear on all parts of the valve operating mechanism and also tends to retard valve opening and advance valve closing. Too little valve lash causes a loss of compression, missing, and eventual burning of the valves and valve seats.

b. **Adjustment Procedure.**

(1) ROTATE ENGINE TO "CLOSED VALVE" POSITION. Open engine grille on left side of tractor and remove distributor cap from distribu-

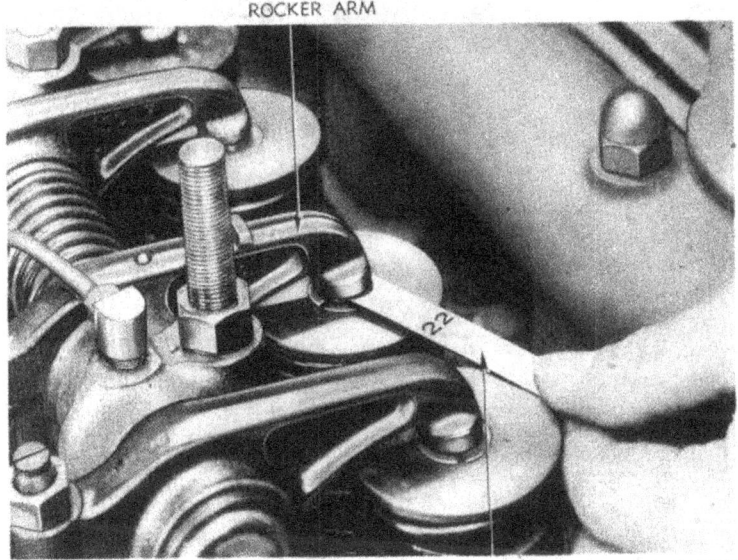

Figure 23—Checking Clearance Between Valve and Rocker Arm

tor (fig. 40). Rotate the engine with wrench on cap screw in end of crankshaft until the rotor arm of distributor corresponds to segment of distributor cap connected by spark plug wire to No. 1 cylinder (closest to rear of tractor) and points are open; then turn engine one-quarter turn in direction of rotation of crankshaft. This will insure both valves of that cylinder being closed.

(2) REMOVE ROCKER ARM COVERS. Remove hand wheel nuts and lift rocker arm covers from engine cylinder heads.

(3) ADJUST VALVES ON REAR CYLINDER. Check clearance between rocker arms and ends of valve stems on both intake and exhaust valves with feeler ribbon (fig. 23). Intake valve should have 0.016-in. clearance, and exhaust valve should have 0.022-in. clearance

111

TM 9-785
58-59

18-TON HIGH SPEED TRACTOR M4

at this point with engine cold. Disregard clearances stamped in plate on side of engine if the clearances given on plate differ from above. Adjustment for this clearance is made by loosening lock nuts on adjusting screws in rocker arms and turning screws in (clockwise) with screwdriver to decrease the clearance, or out (counterclockwise) to increase the clearance (fig. 24). Check again with feeler ribbons after screws are adjusted and lock nuts tightened. There must be a slight drag on 0.016- and 0.022-in. feeler ribbons inserted after

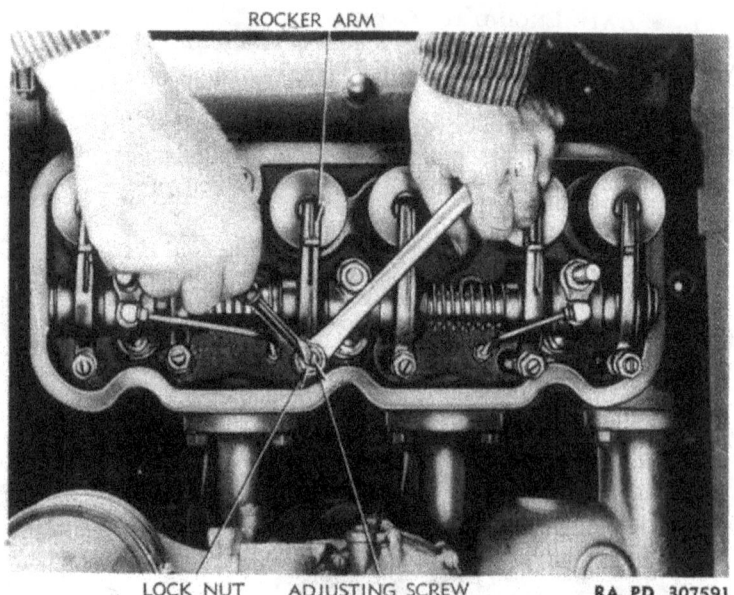

Figure 24—Adjusting Clearance Between Valve and Rocker Arm

lock nuts are tightened. Adjust valves for each remaining cylinder by turning engine each time to close valves on each cylinder as explained in step (1), and using same adjustment procedure.

(4) INSTALL ROCKER ARM COVERS AND DISTRIBUTOR CAP. Install rocker arm covers, being sure gasket is in usable condition, and tighten handwheel nuts. Replace distributor cap in its proper position on distributor. Start engine to see that it operates satisfactorily. Close engine grille.

59. GOVERNOR.

a. **Description.** The governor is of the mechanical non-hunting flyball or flyweight type connected by linkage to the throttle control. Its purpose is to control engine idling speed and to limit maximum engine speed under the variable load requirements.

112

ENGINE

b. Many governors are adjusted unnecessarily because operators fail to realize that irregularities in engine performance are more often due to causes other than faulty governors. A governor, when correctly adjusted, will seldom require attention. The using arm personnel must not attempt any adjustment or replacement of the governor. All other possible causes of irregular engine performance should be eliminated first, and if the governor is still suspected as the cause, report it to the next higher echelon.

60. INTAKE AND EXHAUST MANIFOLDS.

a. **Intake Manifold.**

(1) DESCRIPTION. The intake manifold is of cast iron with jacketing which permits circulation of hot water from the engine over the entire inner casting. A primer manifold is provided and located so that in cold weather, fuel vapor can be injected by the primer pump to aid starting to the fullest extent. An air valve and ignition switch are mounted on the manifold midway between the two carburetors. Both valve and switch are connected by a cable to the "ENG. STOP" knob.

(2) OPERATION. Air and fuel are drawn through the downdraft carburetors and along the water-heated intake manifold where, due to the shape of the air passage and water jacket temperature, the fuel is thoroughly vaporized before it is drawn into the cylinders. When "ENG. STOP" knob is pushed in to stop the engine, the air valve on manifold is opened and ignition switch turned off. With air valve open, air is drawn directly from outlet tube of air cleaner, instead of through the carburetors, and no fuel is drawn into cylinders with the air. In this way engine is forced to stop and the tendency of the engine to continue to run, after ignition is turned off, is overcome.

(3) REMOVAL OF MANIFOLD.

(a) *Disconnect Carburetor Control Linkage.* Remove pins to disconnect upper ends of throttle and governor control rods from throttle shaft levers (fig. 36). Remove pin to disconnect choke rod from carburetor choke control. Remove pin to disconnect control rod from air valve lever (fig. 36).

(b) *Disconnect Air and Fuel Lines.* Disconnect lower ends of flexible fuel lines from Y-fitting on manifold (fig. 25). Remove two cap screws to disconnect Y-fitting from bracket on manifold. Lay fuel line and Y-fitting back out of way. Loosen lower clamp on hose connecting tube from air cleaner to carburetors. Loosen upper clamp on air valve hose and disconnect fuel line from primer manifold fitting (fig. 36). Unhook spring from rear end of throttle shaft.

(c) *Remove Carburetor Assembly.* Remove four nuts and washers from studs holding carburetors to intake manifold and remove carburetor assembly.

18-TON HIGH SPEED TRACTOR M4

(d) *Remove Manifold.* Remove the eighteen nuts and washers from studs holding intake manifold to cylinder heads and remove manifold from tractor. Remove carburetor mounting adapters, air valve assembly, and primer manifold from manifold.

(4) INSTALLATION OF MANIFOLD (refer to figs. 25 and 36).

(a) *Assemble and Install Manifold on Engine.* Install carburetor mounting adapters, air valve and ignition switch assembly, and primer manifold on intake manifold (if they are not already on it). Use new gaskets, unless those removed are in good condition. Shellac gaskets to manifold and install manifold on engine, with a flat washer

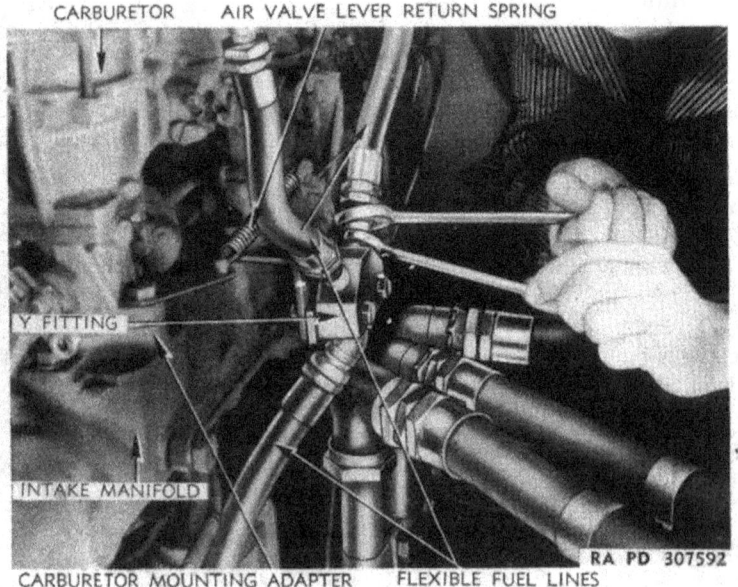

Figure 25—Disconnecting Fuel Line from Y Fitting

under each of the mounting stud nuts. CAUTION: *Draw nuts tight, evenly.*

(b) *Install Carburetor Assembly.* Shellac gaskets to mounting adapters on manifold. Install carburetor assembly on manifold with four nuts and flat washers. Tighten upper clamp on air valve hose.

(c) *Connect Fuel and Air Lines.* Connect end of primer line from pump to front fitting on primer manifold. Secure Y-fitting to bracket on intake manifold and connect flexible fuel lines from carburetors to fitting. Tighten clamp on hose connecting tube from air cleaner to carburetors.

(d) *Connect Engine Controls.* Connect governor and throttle control rods to throttle shaft. Connect choke and engine shut-off con-

ENGINE

trol rods. Check adjustment of controls and adjust as explained in paragraphs 72 c and 74 b.

b. **Exhaust Manifold.**

(1) DESCRIPTION. The exhaust manifold is a three-section iron casting with finned outer surface from which exhaust heat is rapidly dissipated. An exhaust pipe connection is provided at the center section. The front, center and rear manifold sections are counter-bored to receive heavy cast iron ferrules which, when the manifold becomes heated, expand and form exhaust-tight joints.

(2) REMOVAL. Remove four bolts holding elbows to brace on radiators supporting frame between elbows (fig. 49). Remove four bolts connecting lower elbow to manifold and remove elbow. Remove the twelve nuts and flat washers from mounting studs (fig. 26) and lift manifold from engine.

Figure 26—Removing Exhaust Manifold

(3) INSTALLATION. Install exhaust manifold in exact reverse of removal procedure. Use new gaskets between manifold and head and in elbows unless those removed are usable. Tighten all stud nuts evenly to prevent cracking manifold.

61. CYLINDER HEAD GASKET REPLACEMENT.

a. **Remove Intake Manifold Assembly.** Refer to paragraph 60 a (3) for procedure for removal of intake manifold.

b. **Remove Exhaust Manifold.** Refer to paragraph 60 b (2) for exhaust manifold removal procedure.

18-TON HIGH SPEED TRACTOR M4

c. **Remove Water Outlet Manifold.** Loosen clamp on hose connecting the two sections of the manifold. Remove the three acorn nuts and lift section of manifold from head to be removed. Water bypass tube must be disconnected from thermostat housing if removing front section.

d. **Remove Cylinder Head and Gasket.** Remove rocker arm cover, then remove fourteen cylinder head stud nuts and lift cylinder head off studs. Remove gasket from studs. Repeat step c and above operation if both head gaskets are to be replaced.

e. **Install New Gasket and Reassemble.** Scrape all carbon, rust or corrosion from both head and top of cylinder block. Place new gasket or gaskets on studs in cylinder block with side of gasket marked "THIS SIDE UP" facing up (crimped edges around holes in gasket against cylinder block). Set head on studs and install cylinder head stud nuts. Tighten all nuts evenly working from center of head to ends. Use torque wrench and tighten nuts with 175 foot-pounds pull on wrench. Install water manifold on head, using new gaskets. Install intake manifold assembly as explained in paragraph 60 a (4) and exhaust manifold as in paragraph 60 b (3). Adjust valve rocker arms as explained in paragraph 58. Shellac new gasket to rocker arm cover and install cover on head.

62. REMOVAL OF ENGINE.

a. **Remove Ammunition or Cargo Box and Drain Engine Cooling System.** Follow procedure in paragraph 144 b to remove box. Remove plug from end of water drain tube at rear of tractor and open drain valve to drain cooling system. Drain torque converter system as explained in paragraph 116 b.

b. **Remove Engine Hood and Frame.** Remove seven cap screws holding front ends of outer sections of hood to angles at rear of cab. Remove four bolts connecting upper and lower exhaust elbows (fig. 29). Remove two bolts holding generator regulator to hood frame. Remove the generator belt adjusting cap screw from slotted link (fig. 28). Remove two bolts holding front end of angles at sides of center hood section to rear of cab (fig. 27). Remove four bolts holding hood frame to fenders (fig. 28). Fasten rope or chain to hood frame and, using chain hoist, lift off hood, muffler, and water can box as a unit (fig. 29).

c. **Disconnect Fuel, Water, and Air Lines** (fig. 30). Disconnect air lines at rear end of fuel tank. Disconnect water drain line from fitting at bottom of water pump. Remove jam nut from compressor belt idler lubricating tube and slip end of tube out of hole in side frame. Disconnect intake fuel line from fuel pump. Remove adjusting nut and lock nut from air compressor belt idler adjusting screw. Remove the two bolts from rear engine support. Remove large

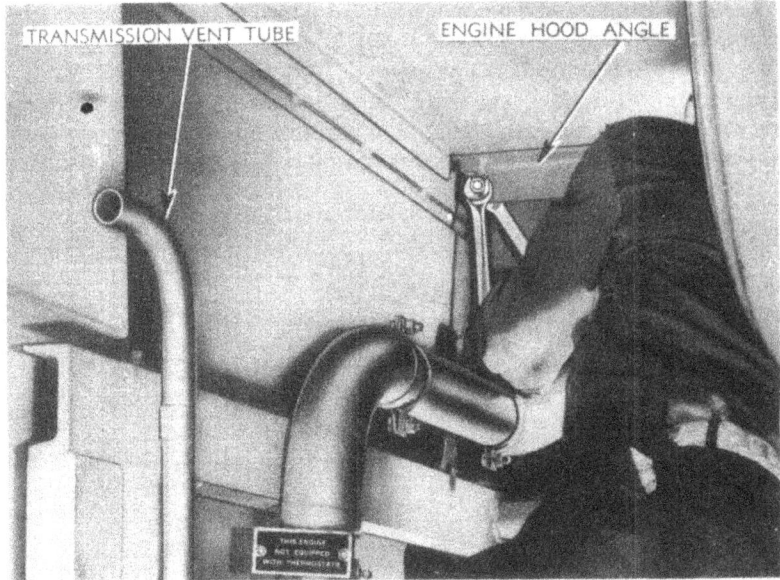

Figure 27 — Disconnecting Hood Frame from Cab

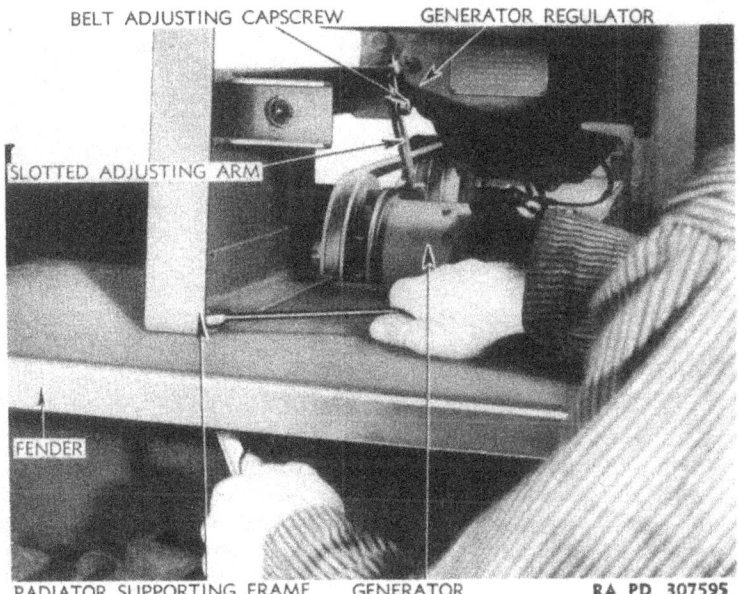

Figure 28 — Disconnecting Hood Frame from Fender

18-TON HIGH SPEED TRACTOR M4

cover below engine from bottom of hull by removing twenty-two cap screws and lock washers (fig. 51).

d. **Remove Radiator Expansion Tank** (fig. 31). Remove cap screws from clip supporting water tube at rear of tank. Disconnect hose connecting tank with radiator at shut-off on tank. Remove four cap screws holding tank to supporting bracket and lift out tank. Re-

Figure 29—Removing Hood, Muffler and Tool Box

move four bolts holding bracket to cab and radiator supporting frame.

e. **Remove Radiator and Supporting Frame.** Disconnect fuel tank vent pipe from fitting at front lower corner of radiator assembly (fig. 32). Disconnect large transmission vent tube at lower end of upright tube. Disconnect converter fluid bleed line from hose back of cab above thermostat housing (fig. 35). Remove all bolts holding supporting frame to hull and fenders (fig. 33). Disconnect wires from the two terminals on generator and disconnect ammeter wire from rear terminal of generator regulator. Loosen clamps on hose connect-

ENGINE

ed to thermostat housing and air intake hose connected to carburetor assembly. Remove two bolts and lift clamp from bracket on fuel tank supporting lower water radiator pipe (fig. 36). Loosen both clamps on hose connecting water radiator pipe to radiator and loosen hose

Figure 30—Rear View of Engine in Tractor

clamp at other end of pipe (fig. 36). Remove the two bolts from clips supporting radiator pipe at rear of cab (fig. 34). Disconnect transmission oil and torque converter fluid lines from radiator connections. Loosen nuts on adjusting screw of fan belt idler on fuel tank and remove belts from fan shaft pulley. Lift off radiator and supporting

TM 9-785
62

18-TON HIGH SPEED TRACTOR M4

Figure 31—Removing Cap Screw from Water Tube Clip

Figure 32—Disconnecting Fuel Tank Vent Line

TM 9-785
62

ENGINE

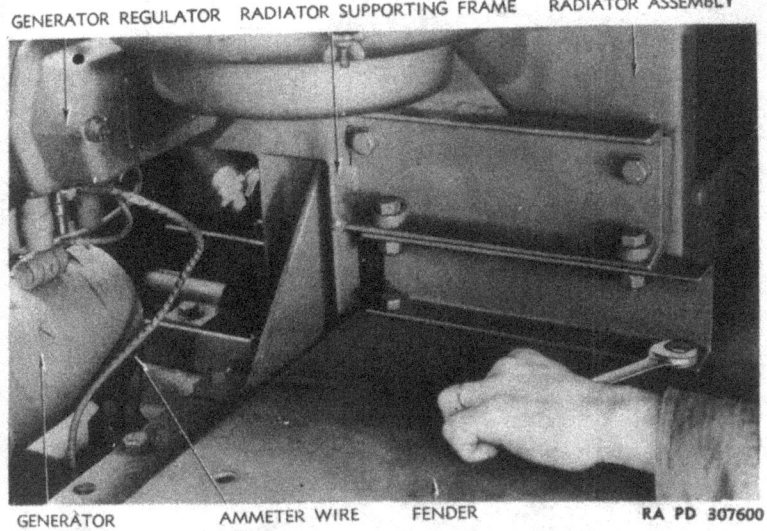

Figure 33—*Disconnecting Radiator Supporting Frame from Fender*

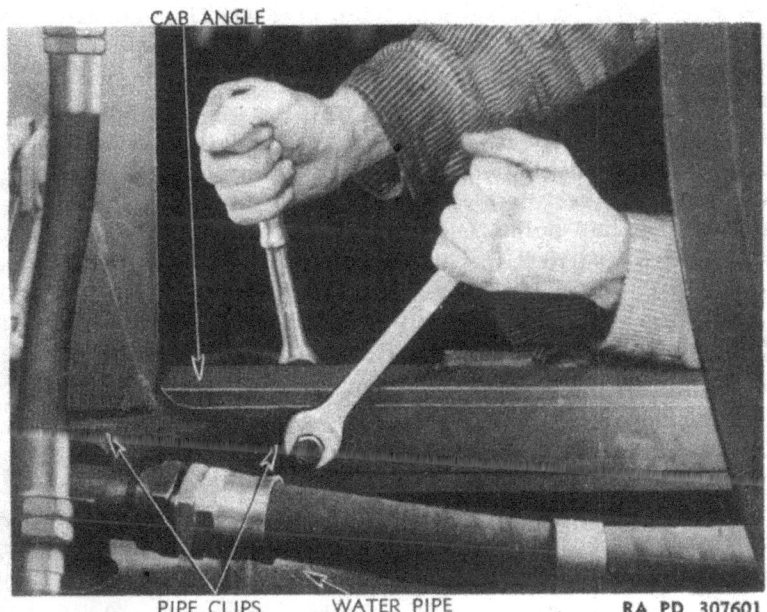

Figure 34—*Removing Bolts from Radiator Pipe Clips*

18-TON HIGH SPEED TRACTOR M4

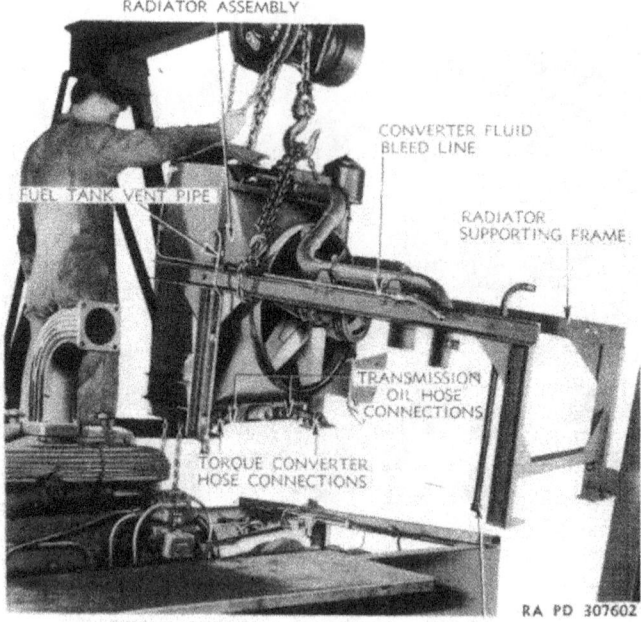

Figure 35—Removing Radiator, Fan, and Air Cleaner Assembly

frame, air cleaner, and fan assembly as one unit with chain hoist (fig. 35).

f. **Disconnect Wires and Engine Controls** (fig. 36). Remove nut and disconnect battery cable from cranking motor (fig. 75). Tape end of cable to prevent sparking by contact with metal. Remove nut and disconnect wire from engine oil pressure gage operating unit near fuel pump. Disconnect wire from ignition switch. Disconnect wire from terminal of thermo gage unit on water manifold. Remove cap screws attaching engine wiring harness clips to cylinder block. Remove pins from yokes to disconnect choke and engine shut-off control cables from engine. Remove nut and disconnect upper end of throttle control rod at ball joint. Remove four bolts supporting radiator cross pipe at rear of cab.

g. **Remove Engine.** Remove nine cap screws connecting engine flywheel housing to clutch housing (fig. 37). Remove lower cap screws working through hull opening with an offset box wrench. Remove two cylinder head stud nuts, one at front of engine and the center one at rear of engine and install the nuts on engine lifter eyes (special) in their place (fig. 38). Tighten nuts firmly, then screw lifter eyes tightly into upper end of nuts. Use lifting sling (41-S-3831-810) as shown in figure 38 with shorter end of bar to rear of engine

A—SPRING
B—GOVERNOR CONTROL ROD
C—CARBURETOR ASSEMBLY
D—CHOKE CONTOLS
E—CHOKE CONTROL CABLE
F—ENGINE SHUT-OFF CONTROL CABLE
G—AIR VALVE LEVER
H—PRIMER FUEL LINE
I—THROTTLE CONTROL ROD
J—PRIMER MANIFOLD
K—WATER RADIATOR PIPE CONNECTING HOSE
L—LOWER WATER RADIATOR PIPE

RA PD 307603

Figure 36—Right Side of Engine in Tractor

and with chain hoist or crane, lift engine slightly, move engine back until clutch shaft is clear of engine, then lift it from tractor (fig. 38).

63. INSTALLATION OF ENGINE.

a. **Place Engine in Tractor.** Aline clutch with pilot bearing in flywheel with clutch arbor if available. Install engine lifter eyes and bar on engine as explained in paragraph 62 g and lift engine into hull of tractor. Maneuver engine back against clutch housing with clutch shaft entering clutch and pilot bearing. Secure flywheel housing to clutch housing with nine cap screws (fig. 37). Install two bolts with lock washers in rear engine support (fig. 30). Install large cover on bottom of hull (fig. 51). NOTE: *Install lock washers on all bolts and cap screws throughout.*

b. **Connect Wires and Engine Controls** (fig. 36). Connect upper end of throttle control rod to ball joint on carburetor throttle lever. Connect throttle and engine shut-off cables. Connect wires to thermo gage operating unit, engine oil pressure gage operating unit

TM 9-785
63

18-TON HIGH SPEED TRACTOR M4

and ignition switch. Install engine wiring harness clips on cylinder block. Connect battery cable to cranking motor (fig. 75).

c. **Install Radiator and Supporting Frame.** Lower radiator supporting frame with radiator assembly, fan assembly, and oil bath air cleaner onto hull over engine (fig. 35). Bolt legs of frame to hull and fenders (fig. 33). Install four bolts through clips on radiator cross pipe above flywheel housing and rear of cab. Connect ends of radiator pipe to radiator and end of radiator cross pipe, then install

RA PD 307604

Figure 37—Removing Cap Screw Connecting Engine to Clutch Housing

the two bolts in clips and rear of cab to support radiator pipe (fig. 34). Tighten hose clamps on radiator pipe connecting hose, then install clamp over pipe and on bracket on fuel tank that supports radiator pipe. Connect transmission oil and torque converter lines to radiator. Connect torque converter, fuel tank, and transmission vent lines. Tighten clamp on hose connected to thermostat housing and on hose connecting air cleaner to carburetor air intake. Install belts on fan drive and fan shaft pulleys and adjust belts for ¾- to 1-inch deflection.

d. **Install Radiator Expansion Tank** (fig. 31). Install four bolts in tank support bracket and cab. Secure tank to bracket with four bolts and connect hose to shut-off on tank. Install cap screw in clip that supports water tube (fig. 31).

TM 9-785
63

ENGINE

e. **Connect Fuel, Water, and Air Lines** (fig. 30). Connect fuel intake line to fuel pump. Connect air lines at rear of fuel tank. Connect water line to fitting at bottom of water pump. Insert adjusting screw of air compressor belt idler through bracket on side of hull and adjust belt for ¾- to 1-inch deflection.

Figure 38—Engine Removed from Tractor with Sling 41-S-3831-810

f. **Install Engine Hood and Frame.** Install hood frame and hood, with tool box and muffler assembled as shown in figure 29. Install four bolts to secure frame to fenders (fig. 28). Install two bolts to connect front end of angles at sides of center hood section to rear of cab (fig. 27). Bolt front ends of outer sections of hood to angles at rear of cab with seven cap screws. Install four bolts to connect exhaust elbows (fig. 49). Install generator belt adjusting cap screw

18-TON HIGH SPEED TRACTOR M4

in slotted arm and adjust belt for ¾- to 1-inch deflection. Install two bolts securing generator regulator to hood frame. Connect ammeter wire to rear terminal of generator regulator (fig. 33), connect wire from front terminal of regulator to "F" terminal on generator and wire from center terminal of regulator to "A" terminal on generator.

g. **Fill Engine Crankcase and Cooling System.** Close engine water drain valve and install plug in end of drain tube. Fill cooling system. Fill engine crankcase with oil to proper level if this has not already been done.

h. **Test Engine and Clutch Control Adjustments.** Check for correct adjustment of choke, throttle and air valve controls as outlined in paragraphs 72 c and 74 b. Test for 1½-inch free travel of clutch pedal (refer to par. 113).

i. **Fill Torque Converter System.** Follow procedure outlined in paragraph 116 b to fill system.

j. **Install Ammunition Box.** Install box as outlined in paragraph 144.

Section XIV

IGNITION SYSTEM

	Paragraph
Description of system	64
Spark plugs	65
Distributor	66
Ignition timing	67

64. DESCRIPTION OF SYSTEM.

a. The ignition system consists of a source of power (generator or battery), the ignition distributor, ignition coil, wiring, and spark plugs. The ignition system, operating through a set of points in the distributor supplies pulsations of direct current to the ignition coil. The coil converts these to high-voltage surges which are produced at the correct intervals and with the correct timing to the engine. Each high-voltage surge produces a spark plug gap which ignites the mixture of air and fuel which had been drawn into the cylinder. When the contact points are closed, current flows through them to the ignition coil, causing a magnetic field to build up in the coil. When the contact points open, the current stops flowing, the magnetic field collapses, causing a high-voltage surge to be induced. This high-voltage surge is led through the wiring, distributor, cap, and rotor, to the correct spark plug.

65. SPARK PLUGS.

a. Spark plugs should be checked carefully with a round feeler gage and set for a 0.025-inch clearance. Wipe all dust and grease from spark plugs and make sure porcelain is not cracked. See that plugs and terminals are tight and that gaskets are not leaking.

  b. Data.
    Make .... Champion No. 6264
    Size 18-mm

66. DISTRIBUTOR.

a. Description. The distributor is a Delco-Remy, Model 1110162 and is a dust-sealed type unit with automatic centrifugal advance. It is located on right side of engine. The distributor shaft and rotor is rotated by a drive operating off the engine camshaft

  b. Data. Clockwise rotation (looking down on distributor)
    Centrifugal advance ...... starts at 400 rpm of engine
    Maximum advance ... 24 degrees at 2,100 rpm of engine
    Point opening .... 0.016—0.018 in.
    Contact point spring tension .... 17-21 ounces
    Cam angle .... 35 degrees

TM 9-785

18-TON HIGH SPEED TRACTOR M4

c. **Adjust Contact Points** (fig. 40). Remove seat back cushion from right-hand end of rear seat. Reach through opening in back of cab, and pull primary wire from center of distributor cap. Pry cap springs off cap and lay cap to one side. Lift rotor and dust seal off shaft. Turn engine until a cam on distributor shaft opens points to widest gap. Inspect points. If they are rough or pitted, smooth them with a small ignition point file so they will make full contact before adjusting (remove distributor to dress points) or replace the points

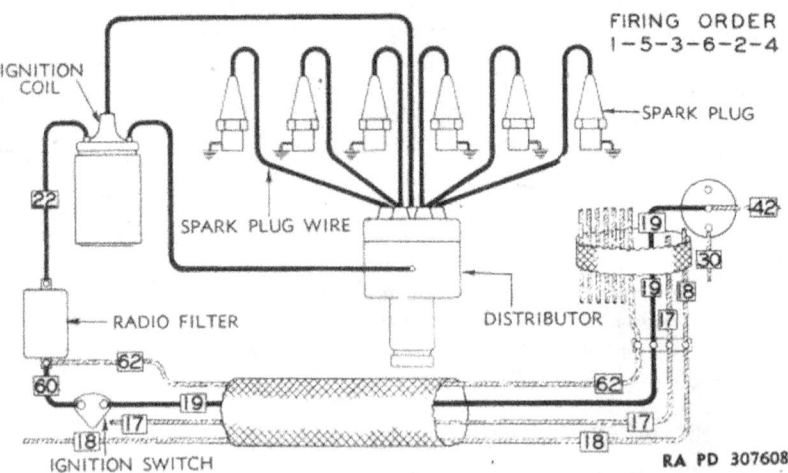

Figure 39—Wiring Diagram—Ignition System

if necessary (see d, e, f, and g). Insert feeler ribbon between the points to check opening gap. If more than 0.018 or less than 0.016 inch, loosen lock screw on stationary point. Turn eccentric screw until an 0.018-inch feeler ribbon will just slip between points without spreading them and a slight "drag" is felt on feeler ribbon. Tighten lock screw and recheck clearance. Correct if necessary. Install dust seal, rotor, and distributor cap, and install seat back cushion.

d. **Remove Distributor.** It is best to first remove distributor from tractor to dress or replace contact points. To remove distributor, follow first part of procedure in c above and remove distributor cap. Then remove cap screw at end of distributor clamping arm, disconnect wire from terminal on side of distributor body, and lift out distributor assembly. If distributor cap is replaced, mark position for No. 1 spark plug wire to be installed in new cap. Every 600 hours, the distributor should be replaced with a new or rebuilt unit. The replaced unit should be sent to a higher echelon for repair or rebuilding.

TM 9-785
66-67

IGNITION SYSTEM

e. **Remove Contact Points.** Loosen breaker lever spring screw, remove breaker lever retaining clip, and lift breaker lever from distributor. Remove both screws from contact support and remove support.

f. **Install Contact Points.** Install contact support in position with two screws. Install breaker lever on post in head, engaging notch of breaker lever spring over spring screw at same time. Install retainer clip and tighten spring screw. Rotate distributor shaft until cam opens points to widest gap and adjust for proper opening as explained in c.

g. **Install Distributor.** Insert end of distributor shaft into drive housing on engine. Install cap screw in end of clamping arm. Install

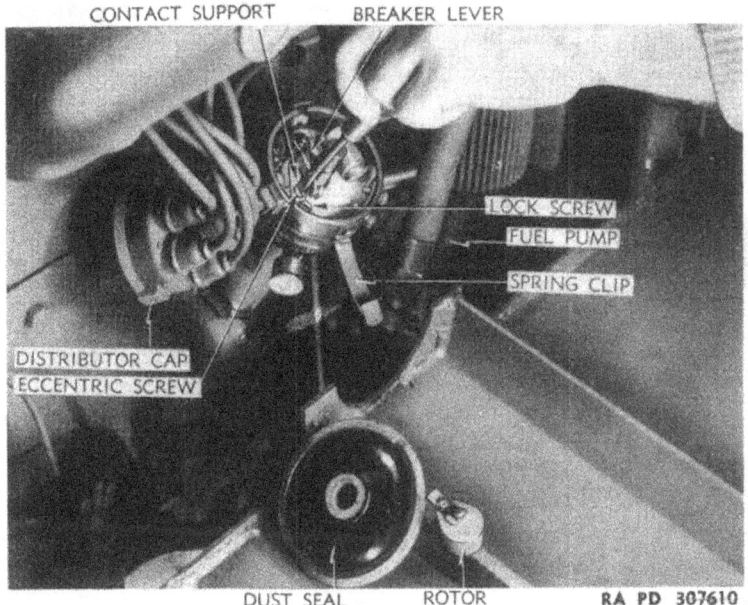

Figure 40—Distributor Point Adjustment

dust seal and rotor on distributor shaft and install cap. Secure cap with spring clips. Insert end of primary wire into center of cap. Connect other wire to terminal on side of distributor. Adjust timing as explained in next paragraph (67).

67. IGNITION TIMING.

a. Accurate ignition timing is important to obtain the maximum efficiency, speed, and power from the engine. Proper timing is obtained by adjusting the distributor so that, with engine running at 2,100 revolutions per minute, ignition will occur at 26 degrees before

129

18-TON HIGH SPEED TRACTOR M4

top dead center with piston nearly at top of its compression stroke. The flywheel is marked for timing No. 1 cylinder (cylinder at end of engine opposite flywheel). There are three marks on flywheel, one "TDC" is at top dead center on flywheel, the second mark, "DIS", is at a point approximately ½ inch or 3 degrees before top dead center, and the third mark is 4½ inches or 26 degrees before top dead center. To time engine, remove small cover from top of flywheel housing (fig. 37). Turn engine until intake valve on No. 1 cylinder closes. Then turn engine slowly until "DIS" mark on flywheel is in line with pointer in flywheel housing. At this point, the distributor contact points should begin to open and rotor should point to segment in distributor cap to which No. 1 spark plug wire is connected. Loosen cap screw on clamping arm and turn distributor so that a 0.0015-inch feeler ribbon can be inserted between points, then tighten cap screw. Recheck point opening after tightening cap screw. With engine running at 2,100 revolutions per minute, the automatic advance mechanism in distributor will have advanced the timing, and ignition should then occur when the third mark, 4½ inches before top dead center passes the pointer. Make this test with timing light according to instructions for use of light. Correct distributor adjustment made with engine stationary, if necessary, for accurate timing under running conditions.

TM 9-785
68-70

Section XV

FUEL SYSTEM

	Paragraph
General	68
Fuel tank	69
Fuel filter	70
Fuel pump	71
Carburetors	72
Primer pump	73
Engine shut-off	74

68. GENERAL.

a. The fuel system considered in this section consists of the fuel supply tank, fuel filter, fuel pump, and carburetors. The fuel is drawn through the fuel filter from the tank and delivered to the dual carburetors by the pump.

69. FUEL TANK.

a. **General.** The fuel tank is located in right side of hull and has a capacity of 125 gallons. The filler tube extends from rear of tank to right rear of hull. One side of tank is recessed for battery mounting. Connections and valves are provided on tank for use of vehicle heaters in cold weather (par. 22). A sediment sump with valve allows water and sediment to settle from tank and be drained out with a minimum waste of fuel. If an ammunition box for 3-inch and 90-mm ammunition is removed from tractor and a cargo box for carrying 155-mm, 8-inch, and 240-mm howitzer ammunition is installed in its place, a filler tube with offset in it (fig. 11) must be installed in place of the straight filler tube in order to allow for opening of door of box to load shells.

b. **Sediment Sump.** Water and sediment must be drained from sediment sump on bottom of fuel tank every week, or daily in freezing weather. Remove small cover from bottom of hull under tank, then open drain cock by turning lever down. Close drain cock again after water and sediment has been drained. CAUTION: *Make sure valve is tight when closed. Install cover.*

70. FUEL FILTER.

a. **Description.** The fuel filter (fig. 41) is mounted on fuel tank below fuel pump. It contains a metal disk element for straining the fuel before it is delivered to carburetors. A drain plug in bottom of filter case provides for draining sediment from the filter.

b. **Filter Service and Replacement** (fig. 41). The small cover on bottom of hull directly under filter must be removed and wheel

TM 9-785
70-71

18-TON HIGH SPEED TRACTOR M4

valve closed before draining filter or removing filter case or element. Remove filter case by removing cap screws in center of filter head. Case and element and spring will then drop down. Wash out filter case with clean fuel, rinse element and blow dirt and cleaning fluid from element with compressed air. CAUTION: *Use care not to damage disks while cleaning.* Use new gaskets when reassembling filter. Open fuel valve and test for leaks after assembly with engine running. Replace cover on hull. If filter is to be replaced, close fuel valve and disconnect fuel lines from filter. Remove mounting cap

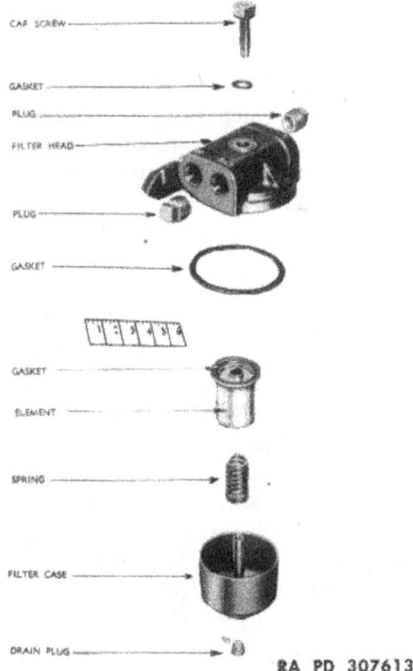

Figure 41—Fuel Filter, Exploded

screws and lift out filter assembly. Install new filter in its place, connect fuel lines and open fuel valve.

71. **FUEL PUMP.**

  a. **Description.** The A-C, Model D-8274, mechanical fuel pump is mounted on side of engine between the fuel tank and carburetors (fig. 40). It draws fuel from the supply tank and pumps it into the carburetor float bowl as it is required by the engine. The pump is operated by a rocker arm on the pump contacting an eccentric on the engine camshaft. Normal fuel pressure is from 3½ to 4 pounds.

FUEL SYSTEM

b. **Replacement of Pump.** Remove seat back cushion from rear of cab. Reach through opening in back of cab and disconnect the suction and discharge lines from pump. Remove the two cap screws holding pump to cylinder block and lift out pump. Install replacement unit on engine, using new gasket and connect fuel lines.

72. CARBURETORS.

a. **Description.** Two Zenith Model 29 downdraft carburetors are mounted on intake manifold and connected to a single air inlet connection. They are of the "set" main jet type with adjustable idling jet. Synchronization of the two carburetors is made possible by the shaft connected to throttle shafts of carburetors. The fuel and air are mixed in the carburetors and metered to the engine in the required amount according to load requirement. The speed of the engine is controlled by the throttle, the variation of amount of fuel delivered to the engine to maintain the desired speed is controlled by the governor, both of which are connected by linkage to throttle shaft levers of carburetors.

b. **Carburetor Replacement.**

(1) If entire carburetor assembly is to be replaced, follow removal procedure outlined in steps (1), (2), and (3) in paragraph 60 c, and steps (2), (3), and (4) in paragraph 60 d for installation. If only one carburetor is to be replaced proceed as in following steps. Due to modifications made after pictures were taken, rods will be found on most engines in place of the choke valve connecting wire and control cables; however, the rods operate in same manner. The air valve lever return spring is also eliminated.

(2) DISCONNECT CONTROLS (fig. 42). Remove pin to disconnect choke control cable from yoke block (or disconnect rod from choke lever) if removing front carburetor. Loosen screws in choke valve levers and pull connecting wire from levers (or disconnect connecting rod). Disconnect hose from Y-block. Disconnect upper end of throttle control rod from carburetor throttle shaft lever at ball joint if removing front carburetor, or upper end of governor control rod from throttle shaft lever if removing rear carburetor. Loosen cap screw clamping connector shaft to carburetor throttle shaft.

(3) REMOVE CARBURETOR. Remove three cap screws connecting carburetor to air inlet connection. Remove nuts from the two carburetor mounting studs. Lift carburetor off studs and slide connector shaft off throttle shaft as carburetor is removed.

(4) INSTALL CARBURETOR AND CONNECT CONTROLS. Use new gaskets between carburetor and air inlet and between carburetor and adapter on intake manifold if old ones are damaged. Reverse procedure of steps (2) and (3) for installation of carburetor. Adjust carburetors and controls as outlined in c.

18-TON HIGH SPEED TRACTOR M4

c. **Carburetor and Control Adjustment.**

(1) ADJUST CHOKE CONTROL (fig. 42). Loosen clamping screws in choke valve levers if wire connects levers. If rod connects levers, disconnect end of rod from one lever. Make sure choke control in instrument panel is in against panel. Make sure springs on choke valve are holding levers against stop. Straighten wire connecting choke valve levers if bent, then tighten clamp screws against wire. If rod connects levers, adjust length of rod by means of adjustable yoke so both valves will be fully open with rod connected. Lubricate linkage if necessary for free movement.

RA PD 307614

Figure 42—Loosening Choke Wire Screw

(2) ADJUST THROTTLE CONTROLS (figs. 42 and 43). Loosen clamp screws in carburetor connecting shaft. Unscrew throttle stop screws on each carburetor until each throttle shaft arm is up against stop pin. Hold both arms fully closed; then tighten connecting shaft clamp screws. Throttle shaft movement will now be the same for both carburetors. Then turn throttle stop screw on front carburetor in (clockwise) against the stop pin to hold the throttle slightly open. Turn stop screw on rear carburetor in even with the first one so both screws will contact their stop pins.

(3) ADJUST IDLING SCREWS (fig. 43). Turn idling adjusting screws in against their seats. Then turn them from 1 to 1½ turns off their seats. Start engine and allow it to reach operating temperature,

FUEL SYSTEM

then close throttle. Idling adjusting screws can then be adjusted for smooth idling of engine. Turning the screws in cuts off air, making the idling mixture richer, turning them out admits more air, making the mixture leaner. If it is necessary to turn the screws in to within less than ½ turn off their seats to obtain good idling of engine, it would indicate either an air leak or a restriction in the flow of fuel for idling. Look for air leaks at the manifold flanges; at carburetor throttle body to intake gaskets, and at carburetor bowl to cover gaskets, due to loosened assembly screws or damaged gaskets. A badly worn throttle shaft will produce sufficient air leakage to

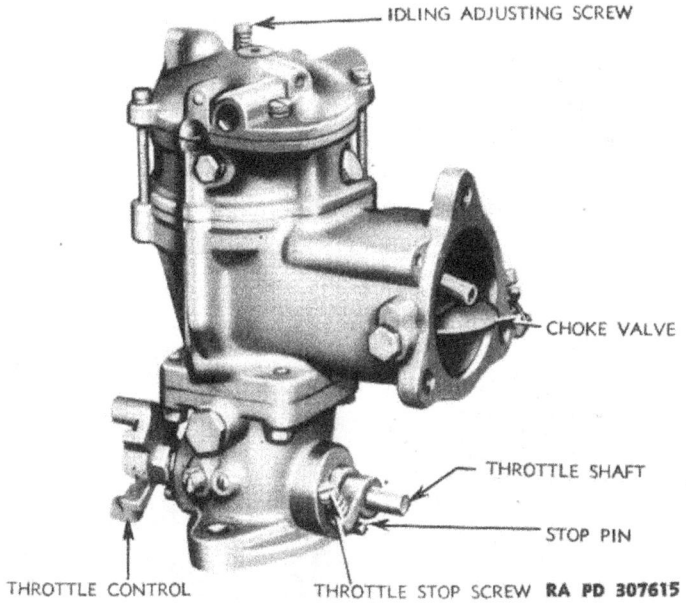

Figure 43—Carburetor Assembly

affect the idling mixture. Dirt or other foreign matter in the idling jet will restrict the flow of fuel for idling and affect the mixture. If the idling jet becomes completely clogged, it will be impossible to run the engine at idling speed regardless of adjustment of the idling adjustment screws.

73. PRIMER PUMP.

a. Description. The primer pump located in instrument panel is a piston type pump with plunger and leathers for drawing fuel through a line connecting it to the fuel tank and pumping it through another line to the primer manifold on intake manifold to aid in starting engine.

TM 9-785
73-74

18-TON HIGH SPEED TRACTOR M4

b. Replacement. Disconnect suction and discharge lines from pump. Hold pump body with wrench or pliers and unscrew nut ahead of pump knob from body of pump. Pull plunger out of pump body, then remove pump body from instrument panel (fig. 44). Install pump by reversing removal procedure.

74. ENGINE SHUT-OFF.

a. Description. The engine shut-off consists of a butterfly type air valve assembly located on intake manifold between the carburetors and ignition switch mounted on a bracket next to shut-off air valve. Both are operated by the same cable from the "ENG. STOP"

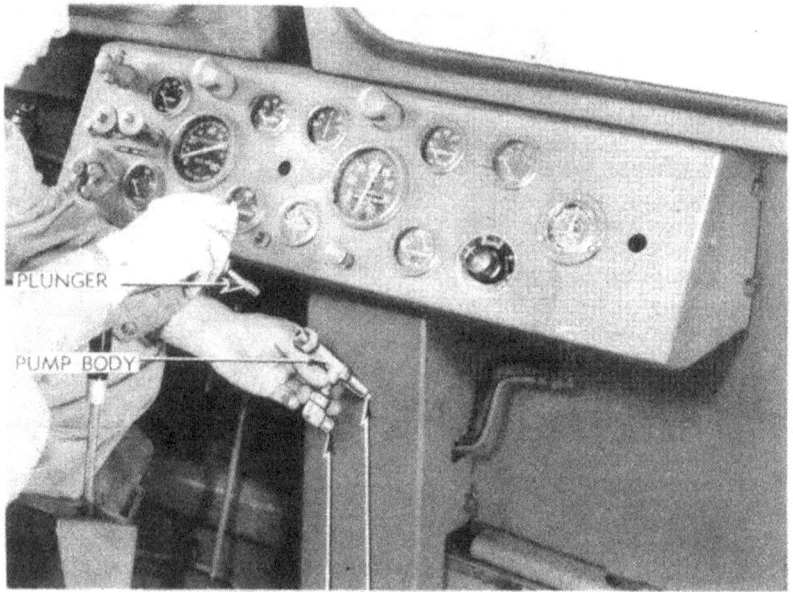

FUEL LINE CONNECTIONS RA PD 307616

Figure 44—Primer Pump Removed

knob on instrument panel. Pulling out on "ENG. STOP" knob opens the air valve and opens ignition switch at same time. The air valve is provided and used in conjunction with the ignition switch to prevent the engine from continuing to run through self-ignition. This is done by the opened air valve allowing the air that is drawn into engine before it stops to pass directly from carburetor air inlet to intake manifold instead of being drawn through carburetors and fuel being mixed with it.

b. Air Valve Adjustment (fig. 45). Air valve is properly adjusted when valve will seat in closed position and slot in valve operating lever on air valve shaft is parallel with intake manifold with

136

TM 9-785
74

FUEL SYSTEM

"ENG. STOP" in running position (pushed in). To adjust, loosen screw clamping wire to operating lever and loosen cap screw clamping lever on air valve shaft. Spread clamp and turn shaft counterclockwise (viewing from right side of engine) until a definite stop is contacted, then turn operating lever on shaft until slot for clamping lever on shaft is parallel with intake manifold. Tighten clamp cap screw. Make sure "ENG. STOP" knob is pushed in against dash, then hold valve tightly closed and tighten screw against wire in lever. On engines with control rod in place of wire connected to valve lever, adjust length of rod for above adjustment.

A—CARBURETOR AIR INLET
B—CONNECTING HOSE
C—IGNITION SWITCH
D—SWITCH LEVER
E—SWITCH BRACKET
F—ENGINE STOP CABLE
G—RETURN SPRING
H—AIR VALVE SHAFT
I—CLAMP SCREW
J—AIR VALVE LEVER
(IN OPEN VALVE POSITION)
K—INTAKE MANIFOLD

RA PD 307617

Figure 45—Engine Shut-off Details

c. **Replacement of Air Valve Assembly** (fig. 45). Loosen screw and pull wire from operating lever. Loosen clamp on hose connecting air valve assembly to carburetor air inlet. Remove front end of return spring from operating lever. Remove two cap screws connecting valve housing to intake manifold and remove valve assembly. Install new valve assembly by reversing above operations.

18-TON HIGH SPEED TRACTOR M4

d. **Replacement of Ignition Switch** (fig. 45). Remove screw to disconnect No. 60 wire from switch. Then remove screw and disconnect No. 19 wire from switch. Tape end of this wire after it is removed. Remove the two mounting screws and lift switch from mounting bracket. Install new switch on bracket, remove tape from end of taped wire and connect it to one of the switch terminals first, then connect second wire.

TM 9-785

Section XVI

INTAKE AND EXHAUST SYSTEMS

	Paragraph
General description	75
Air precleaner	76
Air cleaner (oil bath)	77
Muffler	78

75. GENERAL DESCRIPTION.

a. **Air Intake System.** The air intake system consists of the air precleaner, oil bath air cleaner, and connecting tube (fig. 47). The air drawn from the atmosphere is first drawn through the air precleaner where most of the dirt is trapped, then through the oil bath air cleaner where it passes through oil filled mats and remaining dirt is removed before air passes through the connecting tube to the carburetors. Fuel from the carburetors mixes with the air as it is drawn through the carburetors, making a combustible mixture which is drawn into the cylinders of the engine through the intake manifold and intake valves.

b. **Exhaust System.** After the combustible mixture of air and gasoline is ignited and burned on the power stroke of the pistons,

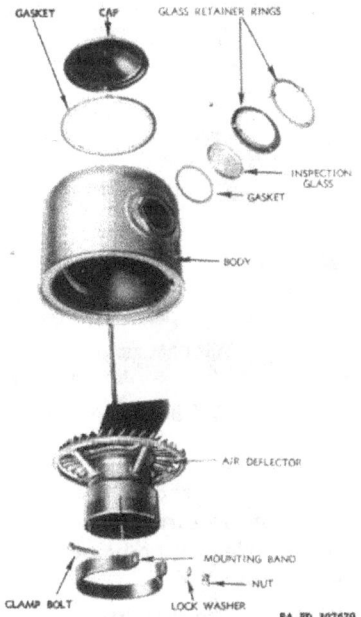

Figure 46—Air Precleaner, Exploded

TM 9-785
75-76

18-TON HIGH SPEED TRACTOR M4

the burned gases are discharged from the cylinders through the exhaust valves into the exhaust manifold. They then pass through the exhaust pipe elbows and muffler (fig. 49) to the atmosphere.

76. AIR PRECLEANER.

a. Description. The cyclone type precleaner is mounted on intake pipe of oil bath air cleaner (fig. 47). As the air is drawn into

Figure 47—Oil Bath Air Cleaner   RA PD 307621

the cleaner, the fins in the cleaner, which are set at an angle, cause the air to whirl and this causes dirt drawn in with the air to be thrown to the outside, and deposited in dirt compartment of cleaner. A glass inspection port on side of cleaner body permits operator to readily determine amount of dirt collected.

b. Precleaner Service. Empty dirt from precleaner daily. If operating under dusty conditions, it will be necessary to empty it

140

INTAKE AND EXHAUST SYSTEMS

oftener (whenever dirt level reaches half-way up on inspection glass). If dirt is allowed to build up higher, it will be sucked over into oil bath air cleaner and would necessitate more frequent cleaning of that unit. CAUTION: *Do not operate tractor with a damaged precleaner.* To empty precleaner, remove wing nut from top cover of cleaner and remove bowl. Shake dust from compartment, wipe glass and reassemble. Replace rubber gasket if present one is not

Figure 48—Removing Oil Bath Air Cleaner

in good condition. Tighten wing nut with fingers. CAUTION: *Do not use a wrench.*

    c. **Replacement.** Loosen bolt in clamp around base of precleaner and lift entire assembly off air cleaner intake pipe. Set new assembly in place, tighten clamp bolt.

TM 9-785
77

18-TON HIGH SPEED TRACTOR M4

77. AIR CLEANER (OIL BATH).

a. Description. The oil bath air cleaner is mounted at side of radiator (fig. 47). Its purpose is to filter all dust from air before the air is delivered to engine. An oil cup with baffle plate is suspended at the lower end of the air cleaner and is filled to a specified level

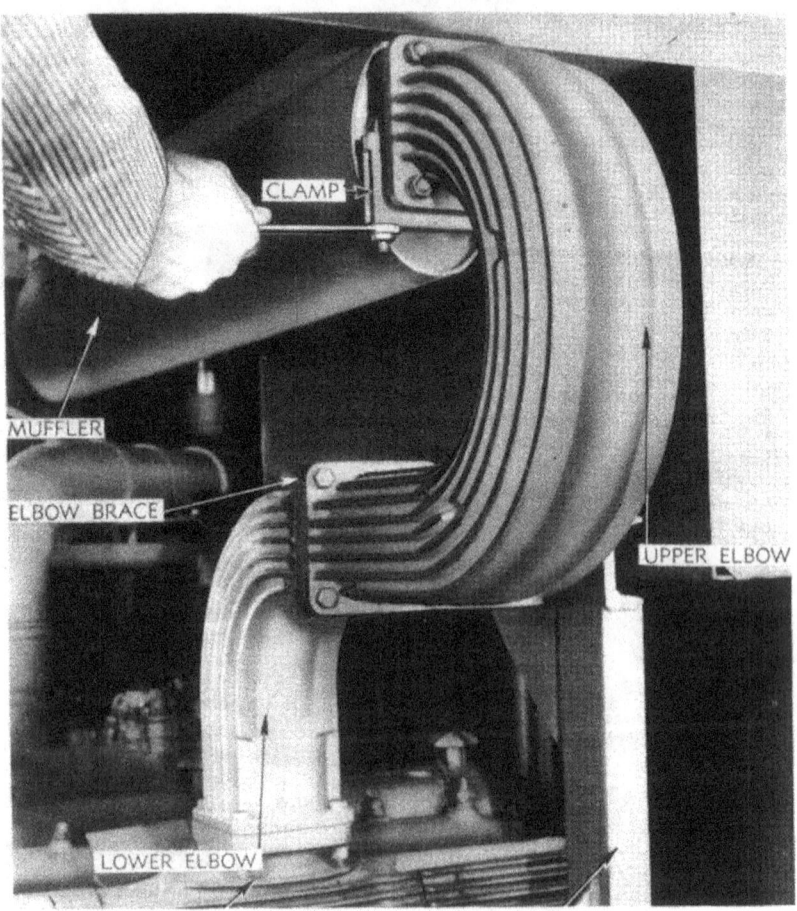

Figure 49—Loosening Muffler Clamp Bolt

with engine oil. As the air is drawn through the cleaner, a portion of this oil is whipped up into screen mats in the main body of the cleaner. The dust in the air collects on these oily screen mats as the air is drawn through them and, as a result, only clean air reaches the engine. The oil dripping back into the cup from the screen mats carries the dirt with it and deposits it in the cup. The cup must be

INTAKE AND EXHAUST SYSTEMS

removed daily, or oftener, and cleaned to remove this dirt. A broken hose, loose clamps, or a leak of any kind between the air cleaner and the engine will defeat the purpose of the cleaner; therefore, care should be taken to see that all connections are tight.

b. Air Cleaner Service. The cleaner must be serviced daily; oftener if tractor is operating under extremely dusty conditions. Remove the oil cup, check oil level in cup and observe the condition of the filtering oil. In extremely dusty conditions, the oil should be changed every eight to ten hours and the air inlet pipe in the air cleaner swabbed out to remove any dust accumulation on the sides of the passage. Improper care of the air cleaner will result in abnormal

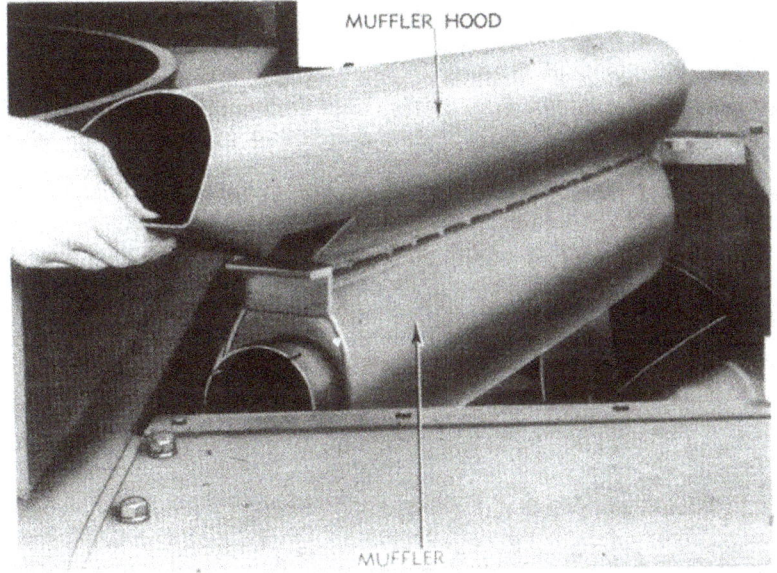

Figure 50—Removing Muffler Assembly

wear on rings, pistons, and cylinder liners. Service as follows: Open radiator grille. Loosen wing nuts holding cup and remove cup. Remove baffle ring and baffle (fig. 47). Empty oil and dirt and thoroughly clean cup, baffle and baffle ring. Replace baffle and baffle ring, and fill cup to the oil level line at top of cone with SAE 30 engine oil when temperature is above 32°F, and SAE 10 engine oil when temperature is below 32°F. CAUTION: *Do not use a Diesel engine oil in the air cleaner, as oils of this type are likely to foam and reduce the efficiency of the air cleaner.* Replace cup on cleaner assembly and tighten wing nut in place around the cup. Close radiator grille.

18-TON HIGH SPEED TRACTOR M4

c. **Replacement of Air Cleaner Assembly.** Open radiator grille. Remove eight cap screws and lift off section of hood over radiator. Loosen clamp on hose connected to air cleaner body and tube to air inlet and disconnect hose from air cleaner. Remove turnbuckle bolt in band around air cleaner body (fig. 48) and lift air cleaner assembly with precleaner attached from tractor. Install replacement unit by reversing removal procedure.

78. **MUFFLER.**

a. **Description.** The muffler is suspended from engine hood frame. Exhaust gases are expelled from the top of muffler through a narrow slot which extends its entire length. A hood with an opening at one end covers this slot. Two large elbows connect muffler to exhaust manifold.

b. **Removal of Muffler Assembly.** Loosen clamp on muffler adapter (fig. 49). Remove bolts from both ends of upper exhaust elbow and remove elbow. Remove crowbar from engine hood section ahead of muffler hood. Remove eight cap screws and lift section of engine hood from tractor. Remove bolts from each end of muffler, swing end of muffler forward (fig. 50), and remove muffler assembly.

c. **Installation of Muffler.** Install muffler by reversing procedure for removal. If old elbow gaskets are damaged, install new ones at muffler and elbow connections.

Section XVII

ENGINE COOLING SYSTEM

	Paragraph
Description of system	79
Filling and draining of system	80
Water pump	81
Thermostat	82
Radiator assembly	83
Cooling fan and fan drive assembly	84

79. DESCRIPTION OF SYSTEM.

a. The cooling system of the engine consists of the water passages in cylinder block and head, water outlet manifold, thermostat assembly, water pump, oil cooler assembly, radiator (center one of the three in radiator assembly), and cooling fan, as well as the necessary water lines for circulation of cooling liquid. The water is circulated by the water pump driven by the timing gears. It draws the cooled water from the radiator and forces it through the oil cooler to cool the oil delivered to the engine. From there it passes into the cylinder block and cylinder head, and out into the water outlet manifold. The thermostat (if tractor is equipped with thermostat) remains closed and water circulates through a bypass tube and through engine only until engine reaches operating temperature. As operating temperature is reached, the thermostat automatically opens to let the water pass into the water outlet manifold, and to the radiator to be cooled. Part of the heated water also circulates through the fuel and air intake manifold to heat the fuel and air as it is drawn into the cylinders. A bypass line in the system provides for circulation of water through the air compressor to cool it. The heated water is delivered to the radiator, and the cooling fan draws air through the radiator, thus dissipating the heat and lowering the temperature of the water while it passes through the radiator from top to bottom.

80. FILLING AND DRAINING OF SYSTEM.

a. **Filling System.** Close drain valve and install plug in end of drain tube at rear of tractor. If possible, fill system with water that is free from lime or alkalines. When filling system or adding water, use clean containers and use care to prevent dirt, sand, or trash from entering filler pipe with water. Tractors may, or may not, have thermostat units removed from thermostat housings depending on climate where tractors are operating. *If units have not been removed*, it is necessary to open valve in thermostat housing to allow air to escape while filling system with water after it has been drained (see b). Use a standard antifreeze solution in the engine cooling system in winter weather. Test solution daily to make sure it is of proper strength for prevailing temperature. Use an antifreeze with

TM 9-785
80-81

18-TON HIGH SPEED TRACTOR M4

a higher boiling point than the normal operating temperature of the engine (160° to 180°F). Any leaks in system must be immediately corrected to prevent loss of water and resultant over-heating of engine.

b. Draining System. To drain engine cooling system, first remove plug from end of drain tube at rear of tractor. Open thermostat valve (only if tractor is equipped with thermostat) (fig. 54) by turning handwheel in counterclockwise direction. This will allow water held in lines above thermostat to drain down. Then open main drain valve by turning valve lever near air compressor (fig. 133) ¼ turn.

81. WATER PUMP.

a. Description. The water pump is of the centrifugal type, located on the timing gear cover and driven by the timing gears. The

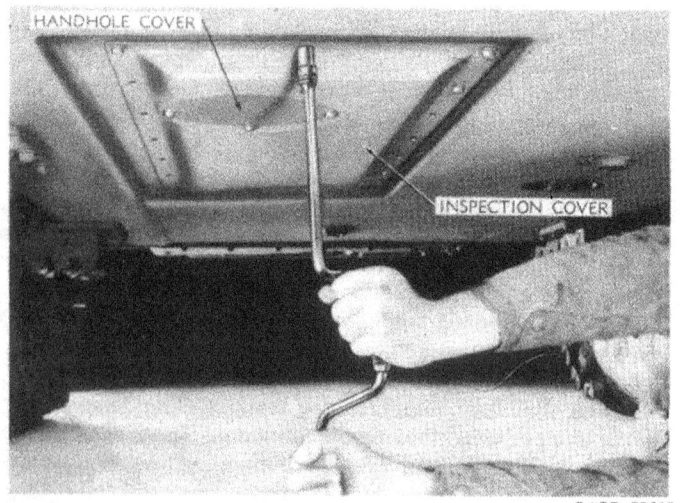

RAPD 59619

Figure 51—Removing Inspection Plate from Below Engine

inlet elbow of the pump connects to the water line leading from lower end of radiator. The discharge end is connected to the oil cooler inlet.

b. Removal of Pump.

(1) OPEN GRILLE AND DRAIN COOLING SYSTEM. Open grille on left side of tractor. Drain water from engine cooling system as explained in paragraph 80 b. If antifreeze solution is being used, drain into clean containers.

(2) REMOVE TOOL BOX. Remove ten bolts holding tool box to left fender and lift off tool box.

146

TM 9-785
81

ENGINE COOLING SYSTEM

(3) REMOVE LOWER INSPECTION PLATE BELOW ENGINE. Remove sixteen cap screws holding plate to hull (fig. 51) and remove inspection plate from tractor.

(4). DISCONNECT PUMP INLET ELBOW. Remove two cap screws from elbow at bottom of pump to disconnect intake elbow from pump. One will be reached from underneath tractor (fig. 52); the other from above. Then remove the two cap screws at front of pump close to cylinder block holding pump to engine.

(5) REMOVE AIR COMPRESSOR. Refer to paragraph 126 c for removal of air compressor.

Figure 52—Removing Cap Screw from Inlet Elbow

(6) REMOVE AIR COMPRESSOR BELT IDLER ASSEMBLY. Remove the two adjusting nuts from belt idler adjusting bolt. Remove two cap screws holding idler bracket to engine and remove idler assembly. Remove the next cap screw below the two just removed. Remove cap screw from rear of engine at top center of pump.

(7) DISCONNECT LINES AND LIFT OUT PUMP. Disconnect oil cooler to bypass valve oil line at valve on side of engine. Loosen water bypass line at both ends and pull line from hoses to clear pump. Lift out water pump assembly.

TM 9-785
81

18-TON HIGH SPEED TRACTOR M4

c. Installation of Pump.

(1) SET PUMP IN PLACE. Cement gaskets to attaching flange on pump and inlet elbow. Lower pump into place, inserting gear into timing gear housing. Insert two long cap screws through pump from front side and top and bottom cap screws from rear side through timing gear housing and into pump attaching flange. Tighten cap screws.

(2) CONNECT INLET ELBOW. Install and tighten two cap screws connecting inlet elbow to inlet of pump.

(3) INSTALL AIR COMPRESSOR BELT IDLER ASSEMBLY. Place idler assembly in position and install two cap screws to attach it to

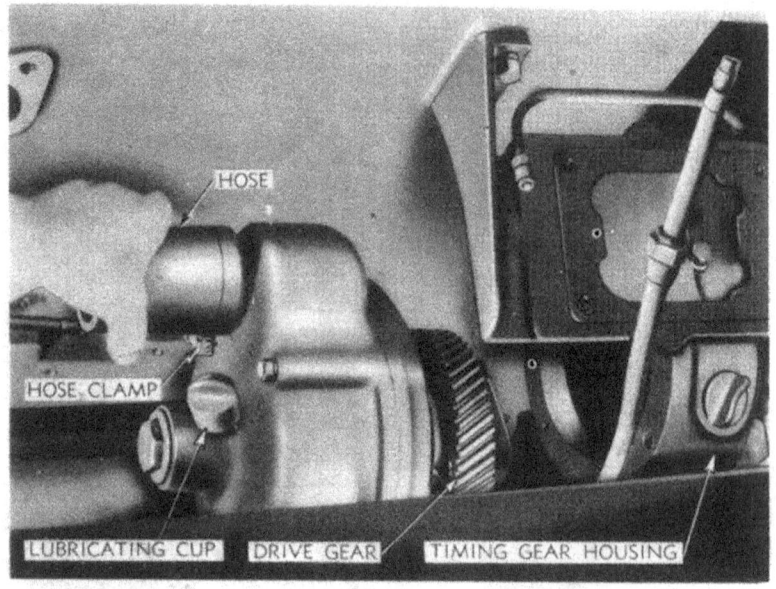

RA PD 59620

Figure 53—Removing Water Pump Assembly

engine. Insert the belt idler adjusting bolt through bracket on hull and start the two adjusting nuts on bolt.

(4) INSTALL OIL AND WATER LINES. Install oil cooler to oil bypass valve oil line, connecting one end to bypass valve on right side of engine and other end to rear connection on oil cooler cover. Shellac inside of ends of hose, and gasket to flange of water bypass line. Slip hose onto pipes and tighten hose clamps. Install two cap screws through intermediate flange on pipe and into cylinder block.

(5) INSTALL AIR COMPRESSOR. Refer to paragraph 126 c for installation of air compressor.

(6) TEST OPERATION OF PUMP. Fill cooling system (par. 80 a), start engine, and check all connections for leaks. Close engine grille.

TM 9-785
82

ENGINE COOLING SYSTEM

82. THERMOSTAT.

a. Description. The thermostat assembly consists of an expansion unit and housing located at the front end of the water outlet manifold. It acts to keep the temperature of the cooling liquid and engine within operating range. When engine is cold, the thermostat is closed and the water circulates through the engine only until the temperature rises. Then the thermostat unit expands and opens, allowing the water to pass to the radiator to be cooled. A valve is provided on thermostat housing which, when opened, allows water to drain from pipes above thermostat when draining system and allows air to escape from engine when filling system. Some tractors are shipped with operating units removed from thermostat housing. A tag

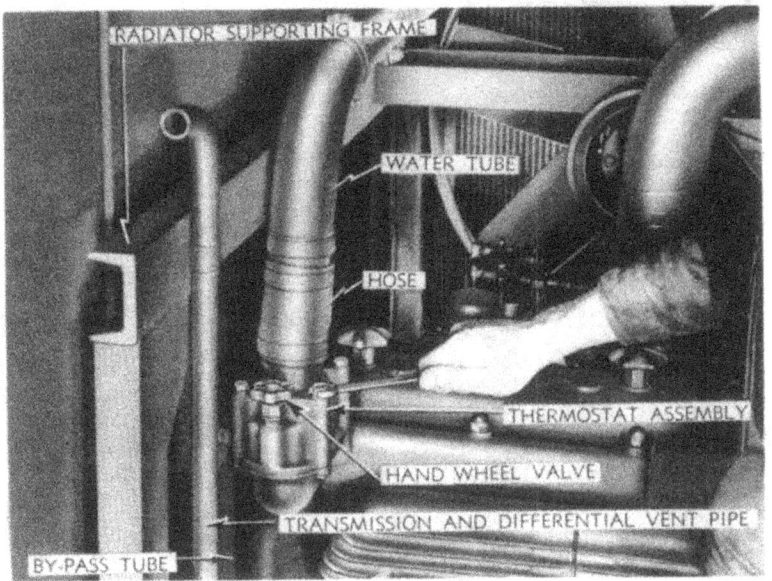

RA PD 307628

Figure 54—Removing Cap Screws from Thermostat Housing

which states "THIS ENGINE NOT EQUIPPED WITH THERMOSTAT" will be found on water tube above thermostat housing on all tractors shipped without them. If tractor is transferred to a cold climate, the thermostat units must be procured and installed so correct engine operating temperature may be maintained.

b. Replacement of Thermostat Assembly (figs. 54 and 55).

(1) DISCONNECT HOSE FROM THERMOSTAT HOUSING. Drain cooling system to level below thermostat. Remove four bolts holding upper radiator pipe brackets to radiator frame above thermostat.

149

TM 9-785
82-83

18-TON HIGH SPEED TRACTOR M4

Loosen clamps on hose connected to top of thermostat housing and hose connected to radiator and raise hose off thermostat housing.

(2) REMOVE THERMOSTAT ASSEMBLY. Remove two cap screws to disconnect bypass tube at front of thermostat. Remove gasket. Remove the four cap screws from thermostat housing and lift off thermostat assembly and attaching gasket.

(3) INSTALL REPLACEMENT THERMOSTAT. Do not attempt to repair thermostat. Place two new thermostats in housing in place of

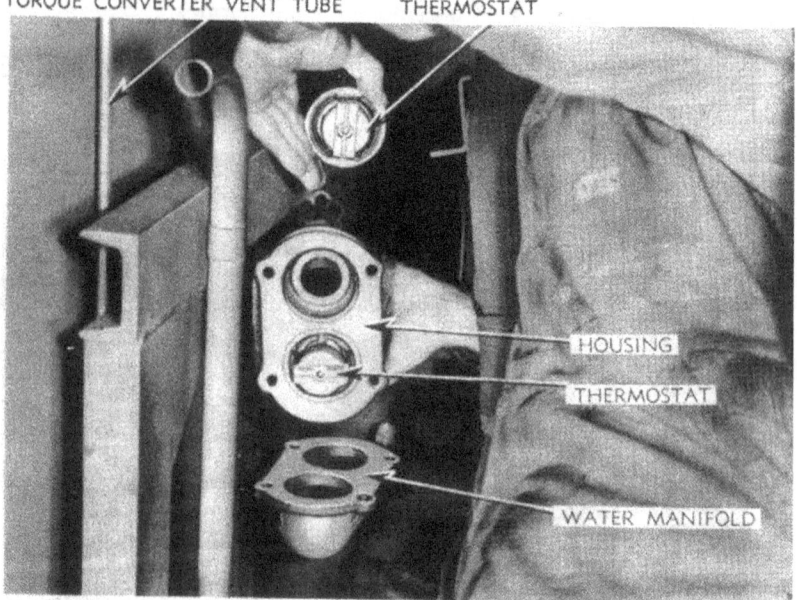

RA PD 307629

Figure 55—Thermostat Assembly Removed

old ones and follow removal steps in reverse order to install. Shellac housing gasket and hose in installation. Do not use shellac on rubber gasket on expansion unit.

83. RADIATOR ASSEMBLY.

a. **Description.** The radiator assembly consists of three radiators of the fin and tube type contained in a saddle. The radiator towards the outside of tractor is for cooling the torque converter fluid,

TM 9-785
83

ENGINE COOLING SYSTEM

the center one is for cooling the water in the engine cooling system and will be referred to as water radiator, and the one closest to the fan is for cooling the oil from the transmission, differential, and power take-off cases. The radiator assembly is supported in a frame extending across the engine. An expansion tank is provided for expansion of water in the water radiator and prevents loss of coolant through expansion and overflow.

b. **Flushing Radiator.**

(1) Flush engine cooling water system periodically to remove accumulated rust or foreign material. This may be done with clean water or, if necessary, with dry-cleaning solvent which is not injurious to steel, cast iron, or copper.

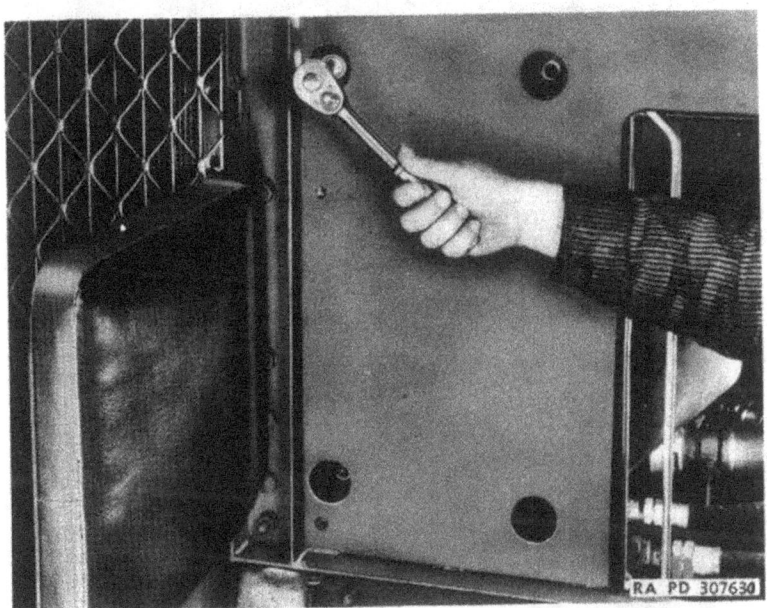

Figure 56—Removing Radiator Mounting Cap Screws

(2) Flush with clean water as follows: Drain cooling water system and disconnect bypass tube from thermostat housing. Remove thermostat unit and install thermostat housing back on engine and connect water bypass tube (par. 82). System may now be flushed. Fill cooling water system, start engine, then open drain valve and, using a hose, keep the radiator filled as the water runs through the system and out. When all the rust, etc., has been flushed from system, stop the engine, install thermostat unit, close drain valve, and refill the system.

151

TM 9-785
83

18-TON HIGH SPEED TRACTOR M4

(3) If a dry-cleaning solvent solution is used to clean the cooling system, a different procedure is necessary. Drain cooling system; then close drain valve and fill system with dry-cleaning solvent. Start engine and run it for about an hour with cover over radiator to hold the engine temperature at 190°F. Then drain the solution from system, flush radiator thoroughly with clean water, and refill system.

(4) If trash or foreign material has gathered at the top of the tubes in radiator, back-flushing of the radiator is necessary. Proceed

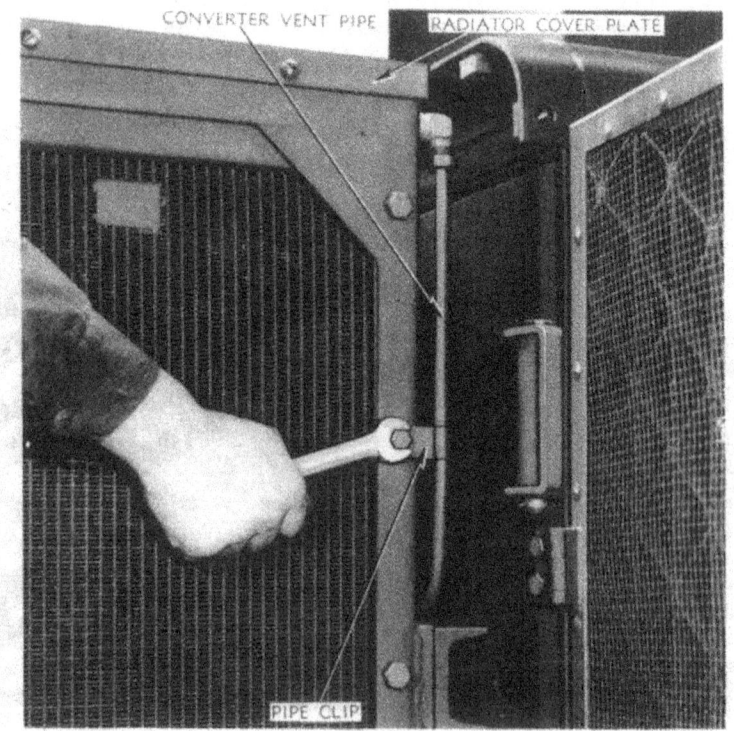

Figure 57—Removing Cap Screw from Converter Fluid Vent Pipe Clip

as follows: Drain cooling system and remove radiator filler cap. Disconnect lower hose of radiator. Insert a water hose inside this hose and stuff a cloth around it if an adapter is not available for connection. Plug opening where hose was disconnected and let water run slowly into radiator through hose until water runs out the top of radiator; then increase the water pressure. Run water through radiator in this way long enough to force the obstructions off the top of

152

ENGINE COOLING SYSTEM

the tubes and out the radiator filler pipe. When finished remove hose and cloth and refill the cooling system.

(5) In extreme cases of clogging, it will be necessary to remove radiator and clean it by other approved methods. Refer to c, d, e, and f for removal and installation procedure.

c. **Remove Seat Back Cushion and Radiator Mounting Cap Screws.** Remove back cushion from right end of rear seat by removing the four cap screws (fig. 56). Remove the four cap screws holding radiator assembly to front side of radiator supporting frame

Figure 58—Radiator Assembly Removed

(fig. 56) and the four cap screws holding radiator to the rear side of supporting frame.

(1) DISCONNECT CONVERTER FLUID, WATER, AND TRANSMISSION OIL LINES. Remove outer section of engine hood above radiator. Remove cap screw from converter vent pipe clip on outer side of radiator (fig. 57) and disconnect vent pipe from elbow at upper end of pipe. Remove radiator cover plate retaining cap screw nearest cab

TM 9-785
83

18-TON HIGH SPEED TRACTOR M4

on inner side of radiator, lift cover plate and disconnect radiator to expansion tank hose from fitting on radiator (fig. 31). Loosen clamps connecting upper radiator hose to upper radiator connection. Reach through opening in back of cab and disconnect converter fluid and transmission oil lines and lower radiator hose from radiator connections (fig. 58).

(2) REMOVE RADIATOR ASSEMBLY FROM TRACTOR. Slide radiator assembly out to edge of fender to clear fan and lines. Install two

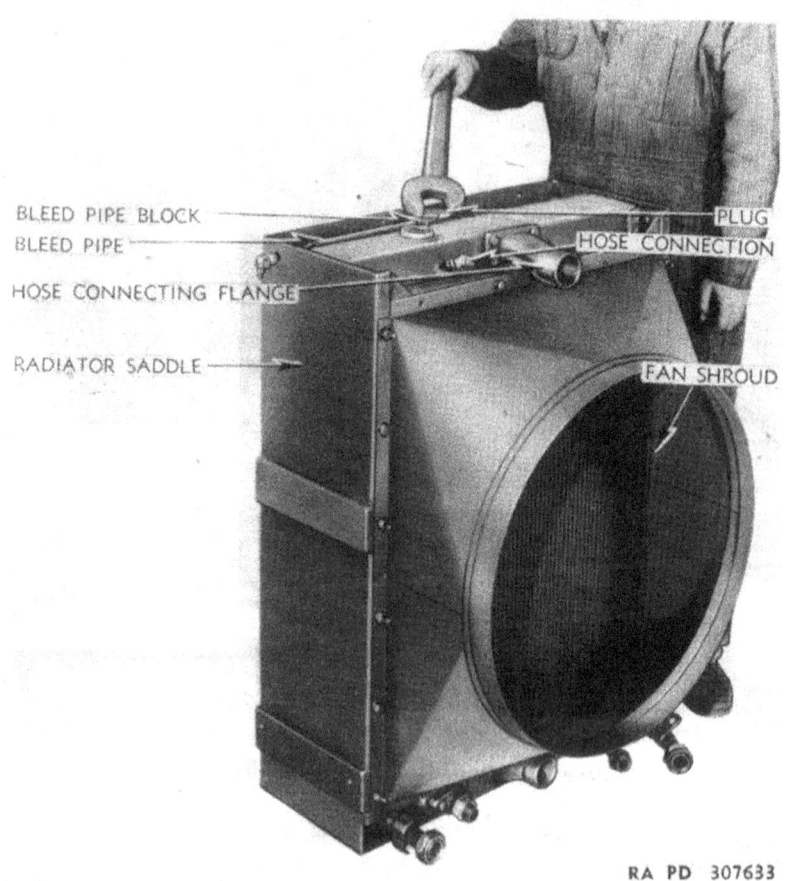

Figure 59—Removing Plug from Torque Converter Fluid Radiator

ENGINE COOLING SYSTEM

cap screws in each side of saddle in holes at upper end, attach rope around assembly under these cap screws, and lift radiator assembly off with chain hoist (fig. 58).

d. **Disassembly of Radiator Assembly.**

(1) REMOVE COVER PLATE, VENT PIPE, AND FAN SHROUD. Remove the two remaining cap screws from inner side of radiator and four screws from outer side and lift off cover plate. Remove large plug

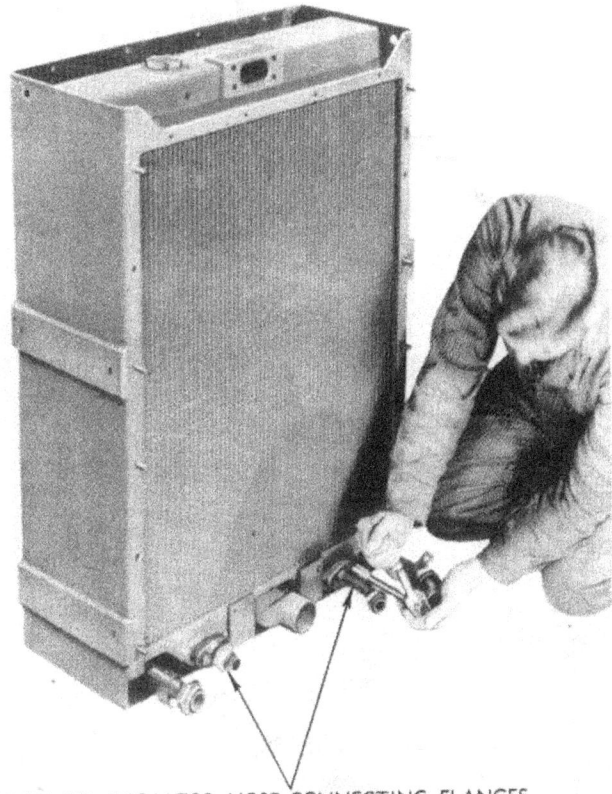

TRANSMISSION OIL RADIATOR HOSE CONNECTING FLANGES
RA PD 307634

Figure 60—Removing Hose Connecting Flanges from Transmission Oil Radiator

and copper gasket from top of converter fluid radiator (fig. 59). Remove copper gasket from below vent pipe block. Unscrew short pipe from block and remove pipe and block. Remove bolts and cap screws from fan shroud and radiator saddle and remove from shroud. Drive long bolts out of radiators and saddle.

TM 9-785
83

18-TON HIGH SPEED TRACTOR M4

Figure 61—Lifting Transmission Oil Radiator from Saddle

(2) REMOVE TRANSMISSION OIL RADIATOR FROM SADDLE. Remove fitting from top of water radiator for connection of hose to expansion tank. Remove four cap screws and remove upper radiator hose connecting flange from water radiator. Remove the two hose connecting flanges from lower end of transmission oil radiator by re-

TM 9-785
83-84

ENGINE COOLING SYSTEM

moving two cap screws from each (fig. 60). Lift transmission oil radiator from saddle (fig. 61).

(3). REMOVE WATER AND TORQUE CONVERTER FLUID RADIATORS. Lift water radiator from saddle. Then remove the four cap screws holding torque converter fluid radiator in saddle and lift out torque converter fluid radiator.

e. **Assembly of Radiator Assembly.** Assemble radiator assembly by reversing procedure for removal in step d. Affix new gaskets, coated on both sides with shellac, to all hose connecting flanges when installing them on radiators.

Figure 62—Removing Cap Screw from Idler

f. **Installation.** Use exact reversal of removal procedure in step e to install radiator in tractor. After installation is completed, fill cooling system as explained in step b and fill torque converter system as outlined in paragraph 116 b (2). Inspect for leaks at connections with engine running.

84. **COOLING FAN AND FAN DRIVE ASSEMBLY.**

a. **Description.** A 6-blade cooling fan of the suction type is used to draw air through the radiators to cool the torque converter fluid, water, and transmission oil as they are circulated through the radiators. Because the fan and radiators are mounted on the side of

157

TM 9-785
84
18-TON HIGH SPEED TRACTOR M4

the engine, a fan drive gear assembly connected by V-belts is used. The fan drive is driven by belts from a pulley on the engine crankshaft, and fan shaft and generator are driven by other belts from the fan drive assembly.

b. **Replacement.**

(1) REMOVE FAN BLADE ASSEMBLY. Open engine grille. Reach over engine and between fan blades and radiator and remove six cap screws holding fan blade assembly onto fan hub. Fan blade assembly may now be removed. If fan shaft assembly is to be removed, drop

Figure 63—Removing Fan Drive Mounting Bolts

fan down into shroud and against radiator until after fan shaft and housing assembly has been removed.

(2) REMOVE FAN SHAFT AND HOUSING ASSEMBLY. Loosen and turn fan belt idler adjusting nuts back to end of bolt. Remove four bolts holding fan shaft assembly to cross member of radiator supporting frame, remove belts from pulley and remove fan shaft and housing assembly.

(3) REMOVE FAN DRIVE GEAR ASSEMBLY. Turn nuts on adjusting bolt of fan drive belt idler at rear of engine back to end of bolt. Remove belts from fan drive pulley. Remove seat back cushion from right side of rear seat. Reach through opening in cab, remove four

ENGINE COOLING SYSTEM

Figure 64—Removing Fan Drive Gear Assembly

cap screws and remove fan belt idler on fuel tank from tractor (fig. 62). Have another man hold weight of drive gear assembly, then remove the four cap screws holding assembly to bracket on side of engine (fig. 63). Remove drive gear assembly through opening in rear of cab (fig. 64).

(4) INSTALL REPLACEMENT UNITS. Install units in reverse order to removal. Adjust belt idlers with adjusting bolts and nuts so that a straight side of belts may be pushed inward ¾ to 1 inch at a point half way between pulleys. Then set jam nuts against adjusting nuts to lock them in place.

18-TON HIGH SPEED TRACTOR M4

Section XVIII

ENGINE LUBRICATING OIL FILTERS AND OIL COOLER

	Paragraph
General	85
Engine lubricating oil filters	86
Engine lubricating oil cooler	87

85. GENERAL.

a. Lubricating oil is delivered to all moving parts of the engine by a gear type oil pump located in the crankcase. The pump draws the oil from the crankcase sump through a screen and delivers it

Figure 65—Removing Element from Filter Case

under pressure to the oil lines and passages in the engine. Before the oil reaches the engine, it passes through an oil cooler element in a housing mounted on the engine just ahead of the water pump. Water from the engine cooling system circulates through the oil cooler housing around the oil cooler element and thus cools the oil. As the oil returns to the sump, part of it is bypassed through two car-

ENGINE LUBRICATING OIL FILTERS AND OIL COOLER

tridge type oil filters where any dirt or sludge that may be in the oil is removed.

b. Pressure of oil delivered to the engine registers on a gage in the instrument panel. Normal pressure at operating engine speed is approximately 40 pounds per square inch. A pressure relief valve is located in the oil pump and opens when pressure exceeds 55 to 60 pounds. This is to prevent stripping gears or excessive wear on gears in pump due to excessive pressure. Another relief valve with an opening pressure of approximately 6 pounds is located at oil cooler inlet which opens to bypass the oil directly to the engine if oil is too

RA PD 307642

Figure 66—Removing Engine Oil Filter Assembly

cold and heavy to circulate through cooler, or in event of oil cooler becoming clogged. A third relief valve is located in right side of engine at the main oil passage through cylinder block. The oil pressure can be regulated to a certain extent with an adjusting screw in this valve assembly, however, all possible causes of abnormal oil pressures should be eliminated before tampering with this adjustment. Report abnormal oil pressure to a higher echelon. The valve in oil pump should be adjusted only in the event pump has been rebuilt. No adjustment is possible on valve in oil cooler inlet.

c. Refer to section XXIII for information on lubrication of transmission, differential, final drives, and power take-off.

TM 9-785
86

18-TON HIGH SPEED TRACTOR M4

86. ENGINE LUBRICATING OIL FILTERS.

a. **Description.** Two filters of the standard military type connected in series are bolted to a bracket on the main frame at the left side of the master clutch housing. They are of the replaceable element type.

b. **Replacement of Filter Elements.** Proceed as follows to replace filter elements: Remove seat cushions from rear compartment of cab. Remove seat plates and raise rear floor plate. Set a small pan under filters, remove drain plugs from filter cases and drain filters. Lift off covers, gaskets, and lift out dirty elements (fig. 65).

Figure 67—Removing Oil Cooler Element

Wash out cases and covers with dry-cleaning solvent, fuel oil, or gasoline; then install and tighten drain plugs. Place new elements in filters. Install covers, using new gaskets. Start engine after tightening cover bolts, and check for leaks. If no leak is evident, lower floor plate and place seat plates and cushions back in tractor. Recheck oil level in engine crankcase, add oil to bring level to "FULL" mark.

c. **Replacement of Filter Assembly** (fig. 66). Remove seats from rear compartment and raise floor plate. Disconnect oil lines from oil hose on filters. Remove four bolts holding filter assembly to bracket and lift out filter assembly. To install filter assembly, reverse removal procedure.

ENGINE LUBRICATING OIL FILTERS AND OIL COOLER

87. ENGINE LUBRICATING OIL COOLER.

a. Description. The oil cooler assembly consists of a housing, inside of which is a metal element somewhat similar to a radiator core (fig. 67). Water from the radiator circulates around the element in the cooler and lubricating oil circulates through the element. The oil cooler plates are lined with small fins which dissipate heat from the oil inside the element to cooling water surrounding the element inside the cooler housing.

Figure 68—Disconnecting Water Line from Oil Cooler

b. Replacement of Oil Cooler Assembly.

(1) REMOVE TOOL BOX. Open engine grille on left side of tractor. Drain cooling system. Open lid of tool box inside grille and remove the ten bolts that hold box to fender. Remove tool box.

(2) REMOVE COOLER ELEMENT. Disconnect inlet and outlet oil lines by removing two cap screws attaching each lower pipe flange to engine (fig. 68). Remove eight cap screws that attach cover to cooler housing and remove cover. Remove element and gaskets. Do not attempt to clean a clogged element. Install a new one.

(3) REMOVE COOLER HOUSING. Disconnect water line from air compressor from fitting on top of oil cooler housing (fig. 68). Loosen

18-TON HIGH SPEED TRACTOR M4

clamp on hose connected to water inlet at rear of housing. Remove two cap screws holding housing to cylinder block, work water inlet from hose, and remove housing.

(4) INSTALL REPLACEMENT UNIT. Install replacement oil cooler assembly in exact reverse procedure to removal. Remove old gaskets and shellac new gaskets onto parts when installing cooler assembly.

Section XIX

ELECTRICAL SYSTEM

	Paragraph
General	88
Battery	89
Generator	90
Generator regulator	91
Starter (cranking motor)	92
Lights	93
Light switches	94
Siren	95
Windshield wipers	96
Wiring system	97

88. GENERAL.

a. Included in this section are all electrical units of the tractor with the exception of the ignition system and instruments. The electrical system is 12-volt throughout with the exception of the blackout driving light and electric brake control units, which are 6-volt. Resistors are employed in the system to reduce the voltage to these units. Electrical current is furnished by a single 12-volt battery. A schematic diagram of the complete electrical system with nomenclature is shown in figures 102 and 103.

89. BATTERY.

a. **Description.** The 12-volt wet cell storage battery is located in a special support built into fuel tank on right side of tractor under floor plate of rear seat compartment. The battery is equipped with handles for easy removal or installation. The negative terminal is grounded to tractor, positive cable connected to a magnetic switch.

b. **Specific Gravity.** Specific gravity readings of each battery should be taken periodically. Use a hydrometer and test each cell separately. If readings are below 1.240, the battery is not receiving sufficient charge. (In zero weather there is danger of freezing if readings are below 1.175; a battery with a specific gravity reading of 1.225 will freeze at 35°F below zero). If water must be added oftener than every two weeks, the electrical system should be adjusted to decrease the charging rate. Otherwise the battery life will be shortened by overcharging.

c. **Hydrometer Readings.** The electrolyte temperature affects the hydrometer reading. For each 30 degrees that the electrolyte is above 77°F, add 10 points to the hydrometer reading to get the true specific gravity. For each 30 degrees that the electrolyte is below

18-TON HIGH SPEED TRACTOR M4

77°F, subtract 10 points from hydrometer reading to get true specific gravity. The above correction is tabulated in the following table for several electrolyte temperatures:

Electrolyte Temperature	Correction to Obtain True Specific Gravity
+122°F	Add 15 points
+107°F	Add 10 points
+ 92°F	Add 5 points
+ 77°F	No Correction
+ 62°F	Subtract 5 points
+ 47°F	Subtract 10 points
+ 32°F	Subtract 15 points
+ 17°F	Subtract 20 points
+ 2°F	Subtract 25 points
− 13°F	Subtract 30 points
− 43°F	Subtract 40 points

Example:
Hydrometer reads 1.250
Electrolyte temperature 17°F
True specific gravity = 1.250 minus 20 points
= 1.230

When taking hydrometer readings on batteries equipped with NO-OVER-FLO, it will be necessary to return all electrolyte withdrawn from battery for purpose of reading by depressing the lead washer. CAUTION: *This washer should be depressed only when returning electrolyte to cell and not when filling with water.*

d. **Adding Water.**

(1) If water is added in freezing temperature and battery is not charged to mix water with electrolyte, the water will remain on top and freeze. In a cold climate, water should be added when the battery is in a room warm enough so that the battery can be sufficiently charged to thoroughly mix the water with electrolyte before the water can freeze.

(2) Distilled water, rain water, or drinking water may be used.

(3) Do not overfill, as subsequent electrolyte expansion may cause flooding and damage. The proper filling height is approximately ⅜ inch above top of separators. Many batteries are equipped with a lead washer in each vent well which is designed to prevent overfilling. Therefore, add water only until it begins to rise into the vent plug well. Draw off excess in order to obtain proper level when vent caps are in place.

TM 9-785
89

ELECTRICAL SYSTEM

e. **Vent Plugs.** Always keep vent plugs in place except when filling or taking gravity readings. Vent plugs must be in place while charging. Be sure hole in vent plug is open.

f. **Keep Batteries Clean and Dry.** If wet or dirty, wash with baking soda solution or ammonia, then with clear water. Be sure vent plugs are tight before washing.

g. **Terminals.** Keep terminals tight and clean. If corroded, disconnect and clean (wash as in step f above). Apply a thin coat of vaseline (not cup grease) to terminal and battery posts before replacing terminal.

Figure 69—Removing Battery Hold-down

h. **Idle Batteries.** An idle battery requires a charge every month or two or at sufficient intervals to keep the gravity above 1.240. Approximate temperatures at which electrolyte will freeze with various specific gravities are as follows:

Actual Specific Gravity	Freezing Temperature
1.270	$-96°F$
1.255	$-60°F$
1.210	$-31°F$
1.185	$-8°F$
1.150	$+5°F$
1.100	$+18°F$

TM 9-785
89-90

18-TON HIGH SPEED TRACTOR M4

i. The battery is in a hard rubber container. When working around the battery, remember that all its exposed metal parts are "alive" and that no metal, tool or wire should be laid across the terminals, as a spark or short circuit will result. Sparks and lighted matches or exposed flames should be avoided near the battery.

j. **Replacement of Battery.** Remove seat cushions and seat plates from rear seat compartment. Lift up rear floor plate and secure with snap fastener. Remove two nuts from hold-down bolts and remove hold-down assembly (fig. 69). Loosen cable clamps, and lift cables from battery terminals. Lift battery out (fig. 70). Install

Figure 70—Lifting Out Battery

replacement battery, reversing removal procedure (also see g). Make sure rubber hold-down blocks on cross bars of battery hold-down assembly are in place and in good condition.

90. GENERATOR.

a. **Description and Data.** The Delco-Remy Model 1105906 generator is a 12-volt, 26-ampere, $5\frac{1}{16}$-inch frame size, ventilated unit, driven by a pulley, with the armature supported at both the drive end and commutator end of the generator by heavy-duty sealed ball bearings. It is mounted on a bracket on fuel tank at rear of tank. The generator output is controlled by the current regulator, in

ELECTRICAL SYSTEM

the manner detailed in paragraph 91. Specifications for generator are as follows:

Cold output—26 amperes at 15 volts at 1,500 revolutions per minute.

Hot output—maximum output controlled by current regulator setting.

Brush spring tension is 22 - 26 ounces.

b. Operation. The operator of the vehicle has the responsibility of observing the manner in which the generator is performing, so that if some abnormal operating condition is noted, proper corrective steps may be taken before complete failure of the equipment takes place. During starting, and while tractor is in operation, the position of the ammeter hand should be noted. If the battery is in a low state of charge, the ammeter hand will indicate a fairly high charging rate. If the battery is in a good state of charge, the voltage control will soon reduce the generator output as the current used in starting is replaced in the battery. This reduction of output is accomplished by the operation of the voltage regulator and results in a tapering off of generator output to a few amperes as the battery reaches a charged condition. This action is indicated by a dropping back of the ammeter needle toward zero. Refer to paragraph 51 if ammeter does not register normal operation.

c. Maintenance. A dirty commutator may be cleaned with No. 00 sandpaper. *Never use emery cloth to clean the commutator.* If the commutator is rough, out-of-round, or has high mica, the generator must be removed so the commutator can be turned down in a lathe and the mica undercut. If the brush length is not sufficient to last until the next inspection period, the brush should be replaced. The pulley nut should be tight, the belt tension should allow for ¾- to 1-inch deflection, and the mounting bolts tight. The connections and wiring in the generator to battery circuit should be in good condition. No lubrication is required.

d. Replacement of Generator.

(1) REMOVE GENERATOR. Open grille on right side of tractor. Loosen cap screw in slotted belt adjusting link (fig. 28) and remove belt from generator drive pulley. Open grille on left side of tractor, reach over engine and remove nuts from "A" and "F" generator terminals. Lift wires from terminals and install nuts again to prevent loss of washers. Remove cap screw connecting belt adjusting link to generator and remove the two generator mounting cap screws. Lift generator from tractor.

(2) INSTALL REPLACEMENT UNIT. Install new unit, placing belt in drive pulley and adjusting for ¾- to 1-inch deflection before tightening mounting cap screws. Connect wire leading from center

TM 9-785
90-91

18-TON HIGH SPEED TRACTOR M4

terminal on regulator to "A" terminal on generator. Generator must now be polarized. Use a jumper lead to momentarily connect "F" terminal to "A" terminal. This allows a momentary flash of current to flow through the field windings which correctly polarizes the generator with respect to the engine. Then connect wire from radio suppression capacitor on generator and wire leading from front terminal on regulator to "F" terminal. CAUTION: *Never operate the*

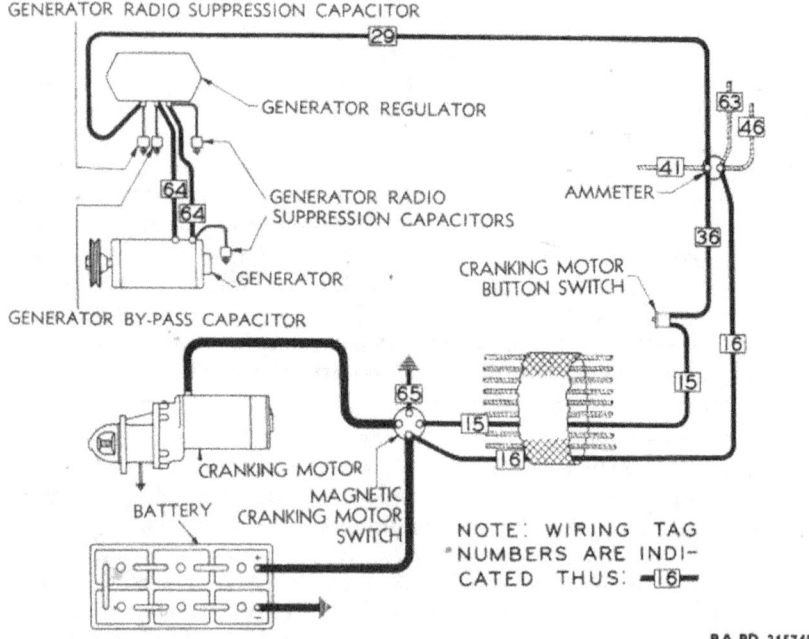

Figure 71—Wiring Diagram—Generating and Starting System

generator with the field circuit connected and the "A" terminal disconnected (open circuit operation) since this would allow high voltage to build up within the generator which would damage the fields and armature.

91. GENERATOR REGULATOR.

a. Description and Data. The Delco-Remy Model 5641 regulator consists of three units, a cut-out relay, voltage regulator, and a current regulator. It is mounted on a bracket at side of radiator (fig. 28).

(1) CUT-OUT RELAY. The cut-out relay closes the circuit between the generator and the battery when the generator voltage has

ELECTRICAL SYSTEM

built up to a value sufficient to force a charge into the battery. The cut-out relay opens the circuit when the generator slows or stops and current begins to flow back from the battery into the generator.

(2) VOLTAGE REGULATOR. The voltage regulator prevents the line voltage from exceeding a predetermined value and thus protects the battery and other electrical units in the system from high voltage. One characteristic of batteries is that as either the specific gravity or the charging rate increases, other conditions being the same, the battery terminal voltage increases. If the terminal voltage is held constant as the battery comes up to charge (specific gravity

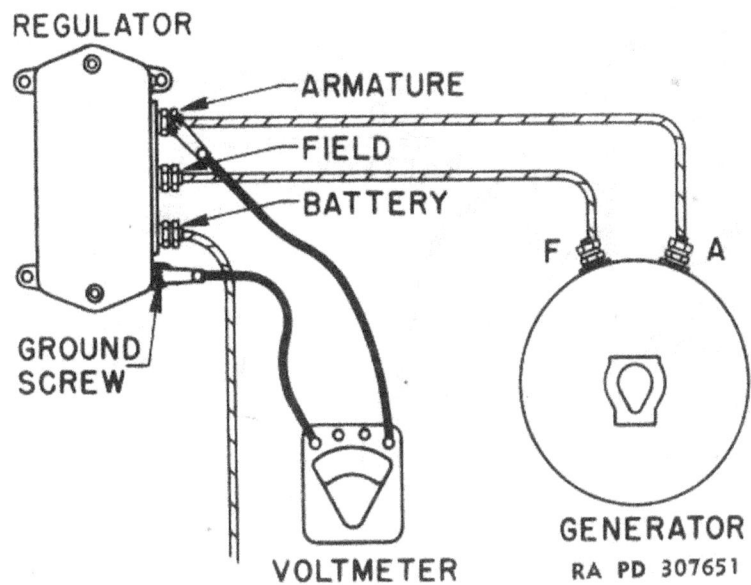

Figure 72—Relay Closing Voltage Check

increases) the charging rate will be reduced. The voltage regulator performs this job of holding the voltage constant and it consequently protects the electrical system from high voltage and the battery from overcharge.

(3) CURRENT REGULATOR. The current regulator limits the generator output to a safe value. It is, in effect, a current-limiting device which operates when the generator output has increased to its safe maximum and it prevents the generator from exceeding this value.

18-TON HIGH SPEED TRACTOR M4

(4) DATA.

Cut-out relay closing voltage............13.5 volts
Air gap ....0.057 in.
Point opening ....0.020 in.
Voltage regulator voltage setting..........15.0 volts
   Point opening ....0.015 in.
Current regulator current setting........25 amperes
   Point opening ....0.015 in.

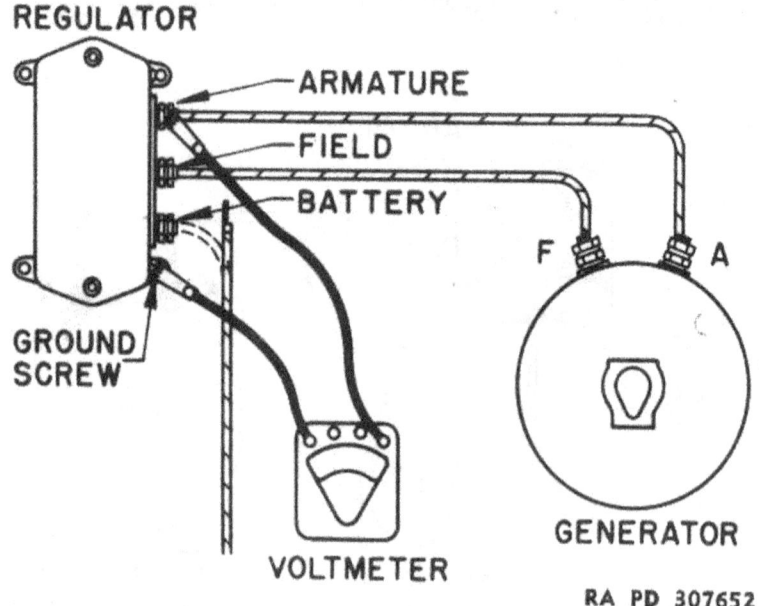

Figure 73—Voltage Regulator Check

   b. **Maintenance.** The electrical settings of the two units may be checked, and replacement made if settings are not in line with the specifications.

   (1) CUT-OUT RELAY CLOSING VOLTAGE. Connect a voltmeter between the armature terminal of the regulator and regulator base (ground screw) as shown in figure 72. Slowly increase generator speed and note closing voltage of the cut-out relay.

   (2) VOLTAGE REGULATOR SETTING. Leave the voltmeter connected between the armature terminal and the ground screw of the regulator as in the previous check, but disconnect the lead from the

ELECTRICAL SYSTEM

regulator battery terminal so the regulator generator system is open circuited (fig. 73). Operate generator at approximately 1,900 revolutions per minute and note voltage regulator setting. Regulator must be hot*—at operating temperature. The check can be made at the end of a run, or the regulator should be operated at least 30 minutes to allow the regulator to attain operating temperature.

(3) CURRENT REGULATOR SETTING. Connect an ammeter into the charging circuit at the regulator battery terminal as shown in figure 74. Two methods of checking the current regulator setting without removing the regulator cover are available.

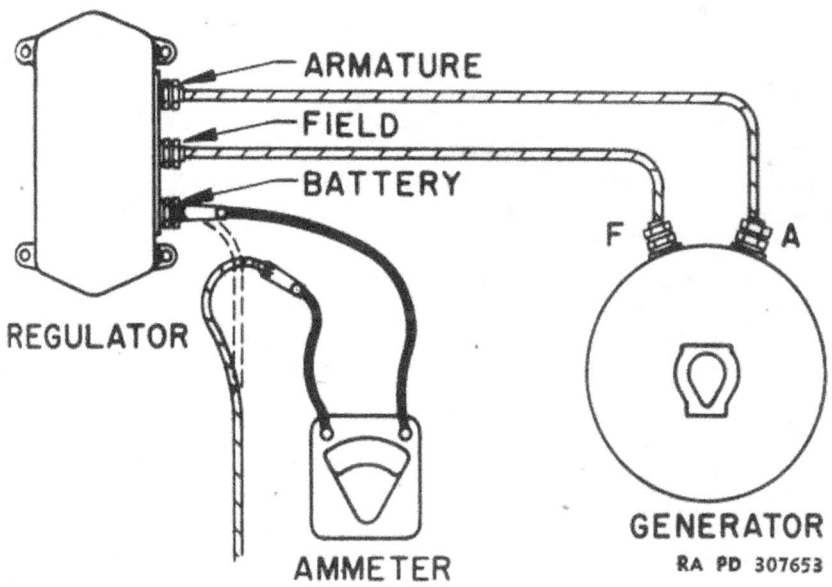

Figure 74—Current Regulator Check

(a) *Battery Discharge Method.* By this method, the battery is partly discharged by cranking the engine for 30 seconds with the ignition turned off so the engine will not start. *Never use the cranking motor for more than 30 seconds at a time without pausing several minutes to permit the cranking motor to cool off.* Immediately after the cranking cycle, start the engine, turn on lights, radio and other accessories so generator output will increase to its maximum as determined by the current setting without causing the voltage regulator

18-TON HIGH SPEED TRACTOR M4

to operate. As current used in starting is replaced in the battery, the battery will come up to charge, the voltage will increase so the voltage regulator begins to operate and tapers off the output. The current regulator setting must be checked before this happens.

(b) Load Method. If a load approximating the current regulator setting (25 amperes) is placed across the battery during the time that the current regulator test is made, the voltage will not increase sufficiently to cause the voltage regulator to operate. This load may be provided by a carbon pile or by a bank of lights.

c. **Replacement of Regulator Assembly.**

(1) REMOVE ASSEMBLY FROM TRACTOR (fig. 28). Open grille on right side of tractor. Remove nuts from regulator terminals and lift wires from terminals. Install nuts and washers back on terminals. Remove the four bolts attaching regulator to bracket and remove assembly from tractor.

(2) INSTALL REPLACEMENT UNIT. Mount regulator on bracket and connect wires as follows: Connect wire from front (generator radio suppression) capacitor on regulator and wire from "F" terminal of generator to front terminal of regulator; connect wire from center (generator bypass) capacitor and wire from "A" terminal of generator to center terminal on regulator; and connect wire from rear (generator radio suppression) capacitor and large wire from ammeter to rear terminal on regulator.

92. STARTER (CRANKING MOTOR).

a. **Description and Data.**

(1) The Delco-Remy Model 644 cranking motor (fig. 75) is mounted in flywheel housing on left side of engine. It is a flange mounted, six-brush, six-pole unit, with Bendix drive, operated by a magnetic switch (par. 94) and using internal reduction gears. The Bendix drive provides automatic meshing of the driving pinion with ring gear on flywheel when the cranking motor button control switch is closed and the magnetic switch is energized. When the engine starts, the drive pinion is automatically disengaged.

(2) DATA. No load—2,000 revolutions per minute at 8.0 volts at 75 amperes
Torque—45 ft-lb torque at 3.5 volts at 500 amperes
Brush spring tension—36-40 ounces

b. **Maintenance.** A dirty commutator may be cleaned with No. 00 sandpaper. *Emery cloth must not be used.* If the commutator is rough, out-of-round, burned, or has high mica, the cranking motor must be removed so the commutator can be turned down in a lathe and the mica undercut. The brushes should be making good clean contact with the commutator, and there should be sufficient brush

TM 9-785
92

ELECTRICAL SYSTEM

length to last until the next inspection period. If the brushes wear rapidly, it may be advisable to remove the cranking motor and check for excessive spring tension, roughness and high mica on the commutator. When the cranking motor is operated, it should take hold promptly, spin the engine at a good cranking speed so that the engine starts. *Never operate the cranking motor more than 30 seconds at a time* without a pause of several minutes, since excessive operation will damage the cranking motor. *Never attempt to move the vehicle with the cranking motor.*

c. **Replacement of Cranking Motor.** Remove seats from rear seat compartment and raise hinged floor plate. Remove nut and lift

RA PD 315746

Figure 75—Disconnecting Cable from Cranking Motor

cable from terminal on cranking motor (fig. 75). Tape end of cable to prevent its touching metal parts of tractor. Remove the three cap screws holding cranking motor in flywheel housing and remove cranking motor through rear seat compartment. Install replacement unit with reverse procedure.

d. **Cranking Motor Switch (Magnetic).**

(1) DESCRIPTION. The magnetic cranking motor switch (fig. 77), mounted on clutch housing, consists of a winding, plunger, contact terminal, and contact disk. When the winding is energized (connected to battery) by the closing of the cranking motor button switch on

175

TM 9-785
18-TON HIGH SPEED TRACTOR M4

instrument panel, the resulting magnetic field pulls in the solenoid plunger, forcing the contact disk against the contact terminals, and connecting the cranking motor to the battery. Opening of the cranking motor button switch disconnects the magnetic switch winding from the battery, so that the magnetic switch spring can separate the contact disk from the terminals, opening the circuit between the cranking motor and battery.

(2) REPLACEMENT OF SWITCH. Remove seats from rear seat compartment and raise floor plate. Remove wires from coil winding terminals (fig. 76) and cables from switch terminals, taping end of each wire as it is removed. Remove switch mounting bolts and re-

Figure 76—Disconnecting Wire from Magnetic Cranking Motor Switch

move switch. Install replacement unit on housing and connect wires to switch as shown in figure 76.

e. Cranking Motor Button Switch (fig. 77).

(1) DESCRIPTION. The cranking motor button switch is a plunger type switch located in instrument panel. When button is pushed in, the circuit between switch terminals is closed and current flows to the winding in magnetic cranking motor switch. When button is released, a spring returns plunger opening circuit to magnetic cranking motor switch.

176

ELECTRICAL SYSTEM

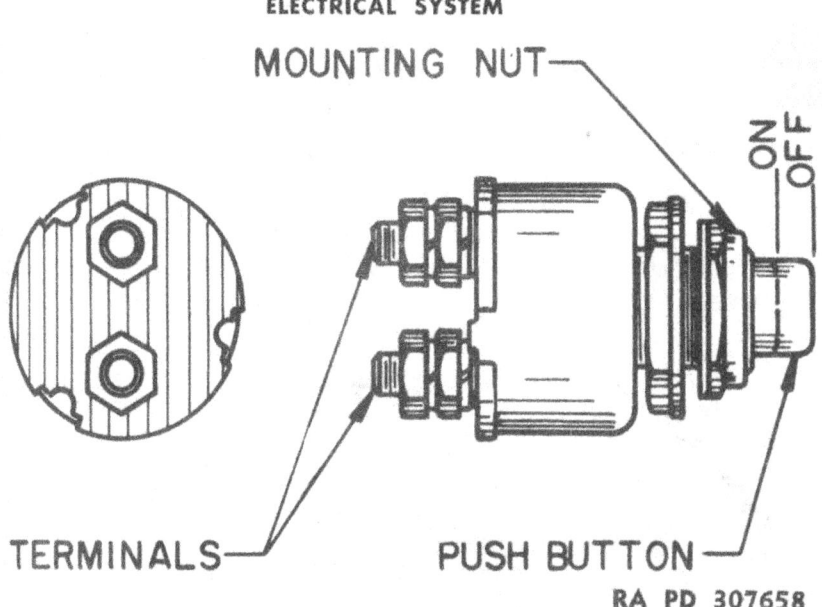

Figure 77—Cranking Motor Button Switch

Figure 78—Disconnecting Wires from Cranking Motor Button Switch

TM 9-785

18-TON HIGH SPEED TRACTOR M4

(2) REPLACEMENT OF SWITCH. Remove mounting nut from switch. Pull switch from panel and disconnect wires from switch terminals (fig. 78). Connect wires to new switch and install switch in panel.

93. LIGHTS.

a. **Description of Lighting System.**

(1) The tractor is equipped with service headlights, blackout driving light and marker lights, floodlights on rear of tractor, combination blackout tail and service stop lights, and panel lights. The

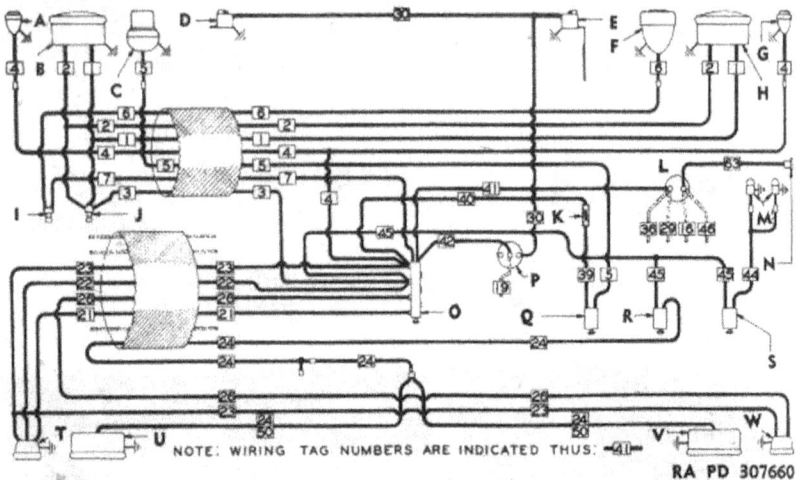

A—LEFT BLACKOUT MARKER LIGHT
B—LEFT HEAD LIGHT
C—BLACKOUT DRIVING LIGHT
D—LEFT WINDSHIELD WIPER
E—RIGHT WINDSHIELD WIPER
F—SIREN
G—RIGHT BLACKOUT MARKER LIGHT
H—RIGHT HEADLIGHT
I—SIREN SWITCH
J—DIMMER SWITCH
K—RESISTANCE COIL
L—AMMETER
M—DASH LIGHTS
N—TROUBLE LAMP SOCKET
O—MAIN SERVICE LIGHT SWITCH
P—WINDSHIELD WIPER SWITCH
Q—BLACKOUT DRIVING LIGHT SWITCH
R—REAR FLOOD LIGHT SWITCH
S—DASH LIGHT SWITCH
T—BLACKOUT TAIL AND SERVICE STOP LIGHT
U—LEFT REAR FLOOD LIGHT
V—RIGHT REAR FLOOD LIGHT
W—BLACKOUT TAIL AND BLACKOUT STOP LIGHT

Figure 79—Wiring Diagram—Lighting System

entire lighting system is 12 volt with the exception of blackout driving light.

b. **Replacement of Headlight or Rear Floodlight.** If removing headlight, remove six bolts and lift off guard. Then pull light from recess in front panel of cab, remove plug from socket in bottom of light by pushing in slightly on plug and turning it to left, then pulling

TM 9-785
93-94

ELECTRICAL SYSTEM

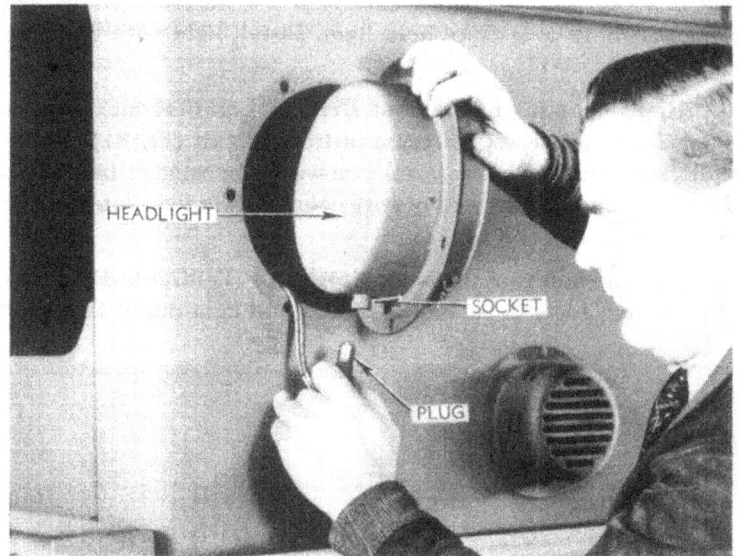

Figure 80—Plug Removed from Headlight Socket

Figure 81—Removing Blackout Driving Light

TM 9-785
93

18-TON HIGH SPEED TRACTOR M4

it out of socket (fig. 80). Rear floodlight is removed in same manner except that only four bolts hold light. Install light with reverse procedure.

c. **Replacement of Blackout Driving Light.** Remove two bolts holding light to bracket in recess in front of cab (fig. 81). Pull light out of recess and pull wire from two-way connector in back of light. Install light by plugging end of wire connected to light into connector, then installing light on bracket.

d. **Replacement of Service Stop and Taillight.** Remove the two nuts from bolts in back of light and pull light out of bracket. Re-

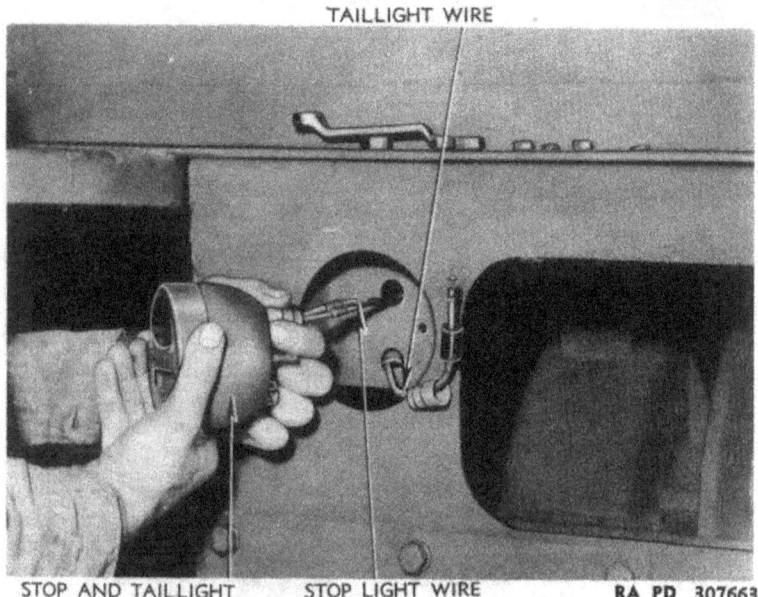

Figure 82—Removing Wires from Stop and Taillight

move wires from back of light by turning plugs to left and pulling plugs from lamp socket (fig. 82). Install by using reverse procedure.

e. **Replacement of Light Bulbs.**

(1) All light bulbs are of the plug-in type held by two lugs on base of bulb which engage in notches in socket with the exception of the blackout driving light. This light is a sealed unit which requires replacement of element when light burns out.

(2) REPLACEMENT OF ELEMENT IN BLACKOUT DRIVING LIGHT. Remove screw and pull off door by pulling out on bottom, then up

180

ELECTRICAL SYSTEM

on door to unhook lug at bottom. Then loosen screws and disconnect wires. Remove clips holding element to door of lamp and remove element. Install new element by using reverse procedure.

(3) REPLACEMENT OF LIGHT BULBS.

(a) Headlight Bulb. Remove headlight guard then loosen screw in clamp ring holding lens. Remove lens, turn bulb and pull it from socket (fig. 83). Install new bulb, then lens and guard.

(b) Service Stop and Taillight Bulbs. Remove two screws and remove lens. Remove bulb from socket and install new bulb. Place lens back on light and install retaining screws.

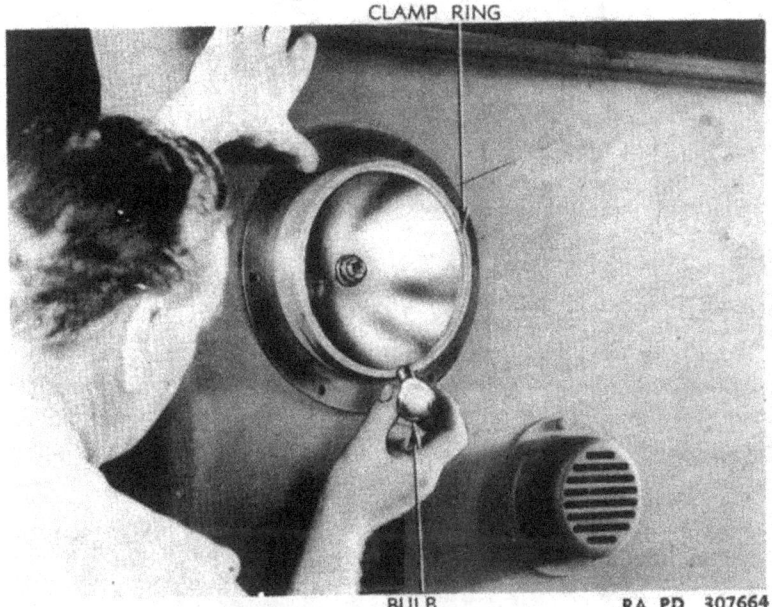

Figure 83—Removing Headlight Bulb

(c) Panel Light Bulbs. Remove two screws and remove shield and glass cover. Remove bulb and install new one. Install glass cover and shield.

94. LIGHT SWITCHES.

a. **General.** The main light switch (fig. 84) controls the service lights, blackout lights, and taillights. This switch has three contacts with a spring stop to lock it in any of its three positions. It may be pulled out to the first position (blackout lights) without depressing the latch button. For the other two positions (service lights and service

18-TON HIGH SPEED TRACTOR M4

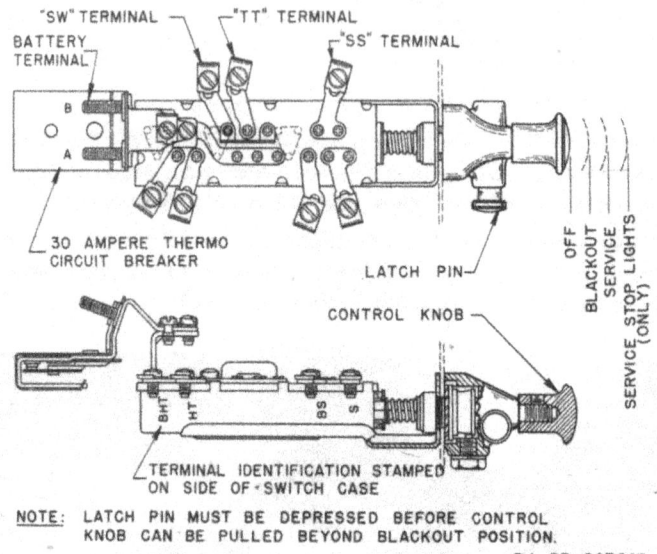

Figure 84—Main Light Switch

Figure 85—Removing Lock Screw from Blackout Control Assembly

TM 9-785
94

ELECTRICAL SYSTEM

stop light), the latch button must be depressed to pull switch out. A foot-operated dimmer switch is provided for dimming lights when meeting other vehicles. Separate switches are provided for blackout driving light, rear floodlights, and panel lights. The main light switch must be pulled out to one of its three positions before any of the other lights can be turned on. There are no fuses in the wiring system for the lights. A thermo circuit breaker on the main light switch, operating in same manner as a thermostat, turns lights off if a short circuit occurs in wiring and prevents burning out of lights. If lights go out and light up intermittently, check for loose connection, broken wire, or bare spot on wire caused by insulation being rubbed off.

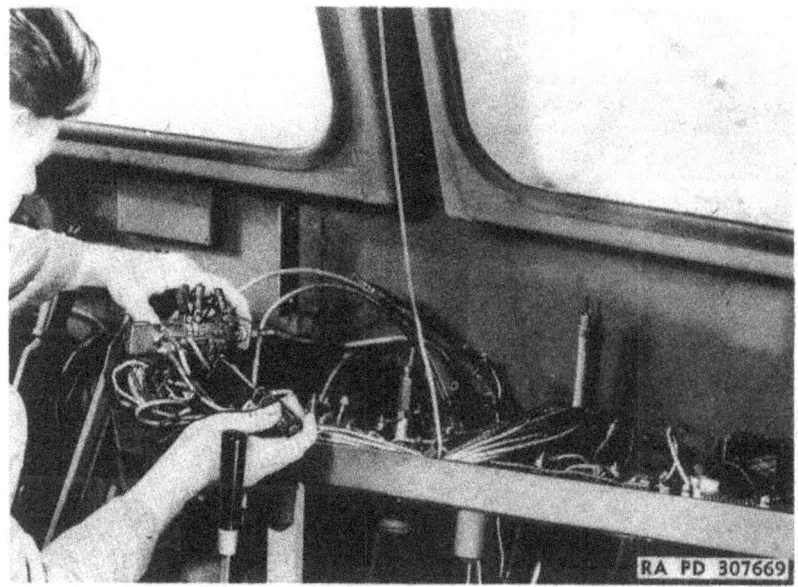

Figure 86—Disconnecting Wires from Light Switch

b. Replacement of Main Light Switch (fig. 84).

(1) DISCONNECT SWITCH FROM INSTRUMENT PANEL. Loosen set screw in control knob and unscrew knob from shaft. Remove lock screw from side of blackout control assembly (fig. 85), depress latch button, and slide assembly from shaft. Remove switch mounting nut.

(2) DISCONNECT INSTRUMENT PANEL FROM DASH. Refer to paragraph 100 b (1) for procedure for disconnecting instrument panel from dash.

183

18-TON HIGH SPEED TRACTOR M4

(3) REMOVE WIRES FROM SWITCH. Pull switch from instrument panel. Remove screws (fig. 86) and nuts and lift wires from terminals of switch. Remove wire leading from ammeter first and tape end of wire. Tag each wire as it is removed with the letters stamped on the switch next to the terminal from which the wire is removed (fig 84). This will insure putting the wires back on the correct terminal when switch is installed.

(4) INSTALL REPLACEMENT SWITCH. Reverse procedure used in removal to install switch. When mounting switch in instrument panel, before tightening mounting nut turn switch so the side to which wires

Figure 87—Removing Wires from Panel Light Switch

are attached faces opposite end of panel. Refer to light wiring diagram (fig. 79) if necessary when connecting wires to switch.

c. **Replacement of Panel Light or Rear Floodlight Switch.** Loosen lock screw in side of knob, remove mounting nut, and pull switch from instrument panel. Disconnect wires from switch (fig. 87), connect them to new switch, and install switch back in instrument panel.

d. **Replacement of Dimmer Switch.** Remove bolts and remove switch from bracket. Disconnect wires from switch (fig. 88). Install

ELECTRICAL SYSTEM

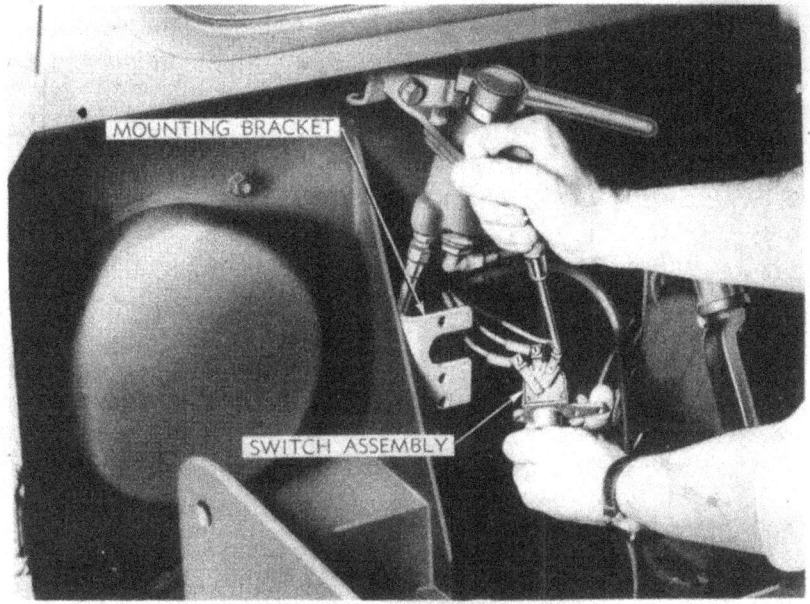

Figure 88—Disconnecting Wires from Dimmer Switch

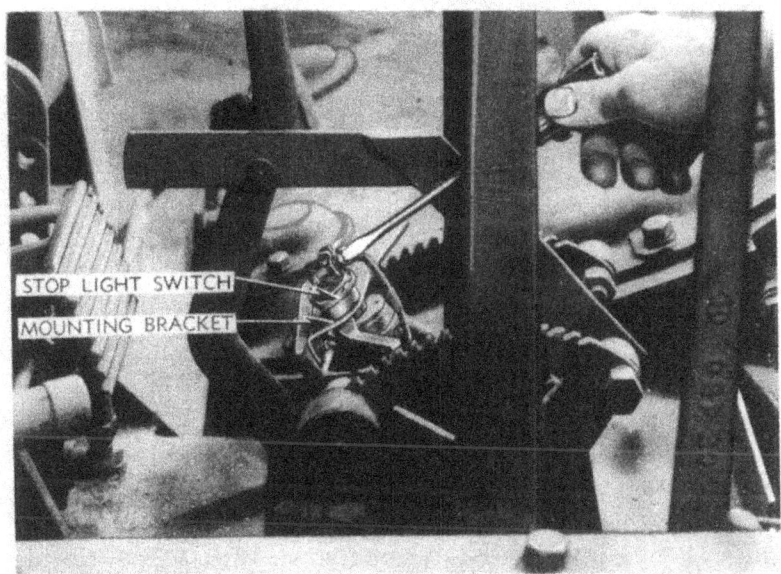

Figure 89—Disconnecting Wires from Stop Light Switch

TM 9-785
18-TON HIGH SPEED TRACTOR M4

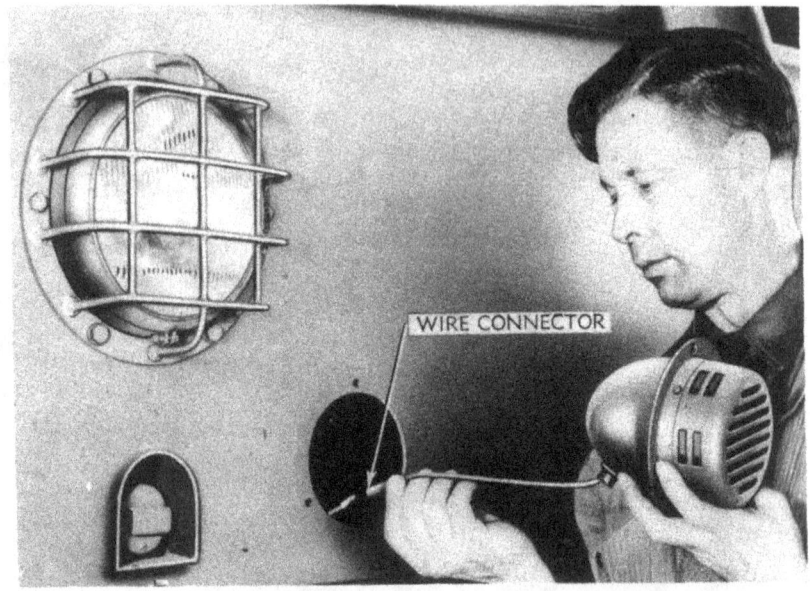

Figure 90—Disconnecting Siren Wire

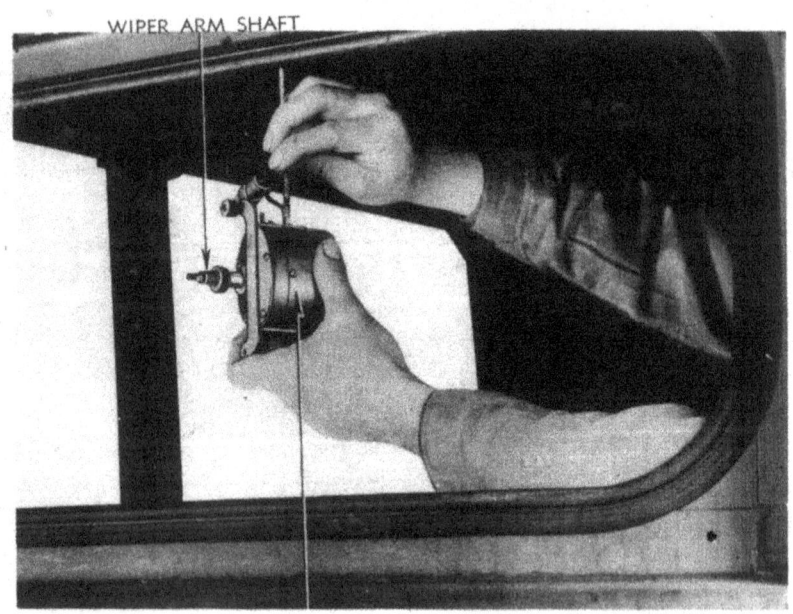

Figure 91—Removing Windshield Wiper Motor

ELECTRICAL SYSTEM

each wire on corresponding terminal of new switch as it is removed from terminal of switch being removed.

e. **Replacement of Stop Light Switch.** Remove screws and lift the two wires from stop light switch terminals (fig. 89). Remove two screws holding switch to bracket and remove switch. Install new switch on bracket and connect the wires to switch terminals.

95. SIREN.

a. **Description.** The siren is of the standard military type and mounted in front panel of cab. It is operated by a switch similar to the dimmer switch for the headlights.

b. **Replacement of Siren.** Remove three bolts holding siren to front panel of cab. Pull siren from recess in cab, disconnect wire from connector as siren is removed (fig. 90). Insert wire of replacement siren into connector and install siren.

96. WINDSHIELD WIPERS.

a. **Description.** The two windshield wipers are driven by separate electric motors controlled by a single switch. The two wipers work in unison when switch is turned on.

b. **Replacement.** Loosen screw nearest wiper motor shaft on wiper arm and pull wiper arm from shaft. Remove shaft nut and pull arm drive from shaft. Remove nut from shaft housing and remove two bolts holding wiper motor to cab. Pull motor from cab frame, pull wire from motor, and remove motor (fig. 91). Use reverse procedure for installing replacement unit. When installing wiper arm on shaft adjust it so that arm swings an equal distance both ways.

97. WIRING SYSTEM.

a. **Description.** Figures 102 and 103 are schematic drawings of the complete wiring system of the tractor. The various electrical units and accessories are shown in their relative locations when installed in tractor and are designated by number just which wire or wires connects to each. All the wires on the tractor have metal tags wrapped around them at each end with numbers stamped on the tags corresponding to the numbers on the wires in the drawing with the exception of the spark plug ignition wires and two wires on the hour meter. Some of the wires are made up in harnesses while others are single. The entire system can be easily wired by following this diagram.

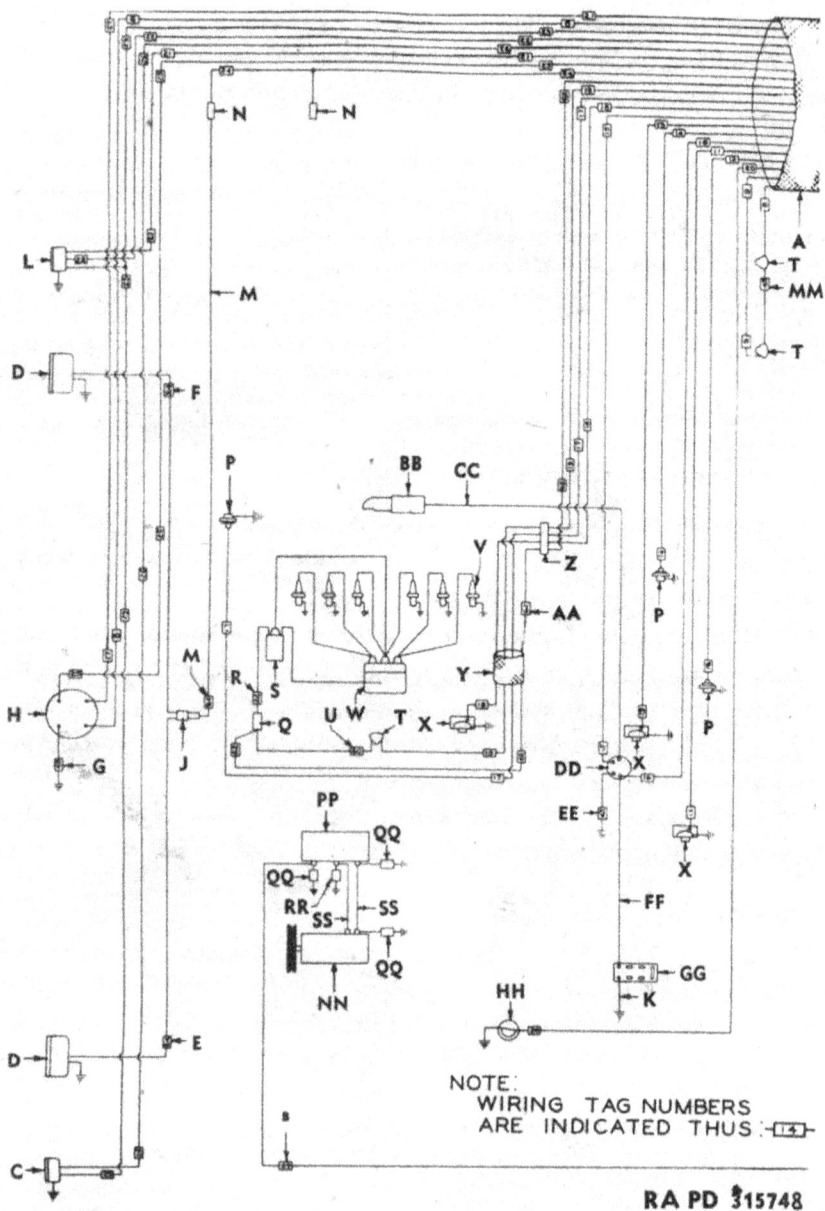

Figure 92—Wiring Diagram—Rear Part of Tractor

ELECTRICAL SYSTEM

A—MAIN WIRING HARNESS (TAG NOS 8 THRU 27 AND 62)
B—RADIO FILTER TO AMMETER WIRE, NEGATIVE (TAG NO 29)
C—BLACKOUT TAIL AND BLACKOUT STOP LAMP
D—REAR FLOOD LAMP
E—REAR FLOOD LAMP WIRE (TAG NO. 24)
F—REAR FLOOD LAMP WIRE (TAG NO. 24)
G—GROUND TO COUPLING SOCKET WIRE (TAG NO. 48)
H—ELECTRIC BRAKE COUPLING SOCKET
K—BATTERY CABLE (NEGATIVE TO GROUND)
L—BLACKOUT TAIL AND SERVICE STOP LAMP
M—REAR FLOOD LAMP WIRE, FRONT SECTION (TAG NO. 24)
N—WIRING CONNECTOR (2-LINE)
P—TEMPERATURE GAGE UNITS (ENGINE, TORQUE CONVERTER AND TRANSMISSION)
Q—IGNITION COIL RADIO SUPPRESSION FILTER
R—RADIO FILTER TO IGNITION COIL WIRE, PRIMARY (TAG NO. 22)
S—IGNITION COIL
T—IGNITION SWITCH
U—IGNITION SWITCH TO RADIO FILTER WIRE (TAG NO. 60)
V—SPARK PLUG
W—IGNITION DISTRIBUTOR
X—OIL PRESSURE GAGE UNITS (TORQUE CONVERTER, TRANSMISSION AND ENGINE)
Y—ENGINE WIRING HARNESS (TAG NOS. 17-18-19 AND 62)
Z—WIRING JUNCTION BLOCK IGNITION
AA—JUNCTION BLOCK TO RADIO FILTER WIRE (TAG NO. 62)
BB—CRANKING MOTOR
CC—BATTERY CABLE (MAGNETIC SWITCH TO CRANKING MOTOR)
DD—CRANKING MOTOR MAGNETIC SWITCH
EE—GROUND TO MAGNETIC SWITCH WIRE (TAG NO. 65)
FF—BATTERY CABLE (POSITIVE TO MAGNETIC SWITCH)
GG—STORAGE BATTERY
HH—FUEL TANK GAGE UNIT
MM—STOP LAMP SWITCH WIRE (TAG NO. 49)
NN—GENERATOR
PP—VOLTAGE AND CURRENT REGULATOR
QQ—GENERATOR RADIO SUPPRESSION CAPACITORS
RR—GENERATOR BY-PASS CAPACITOR
SS—GENERATOR TO REGULATOR WIRE (TAG NO. 64)

RA PD 315748B

Legend for Figure 92—Wiring Diagram—Rear Part of Tractor

TM 9-785
94

18-TON HIGH SPEED TRACTOR M4

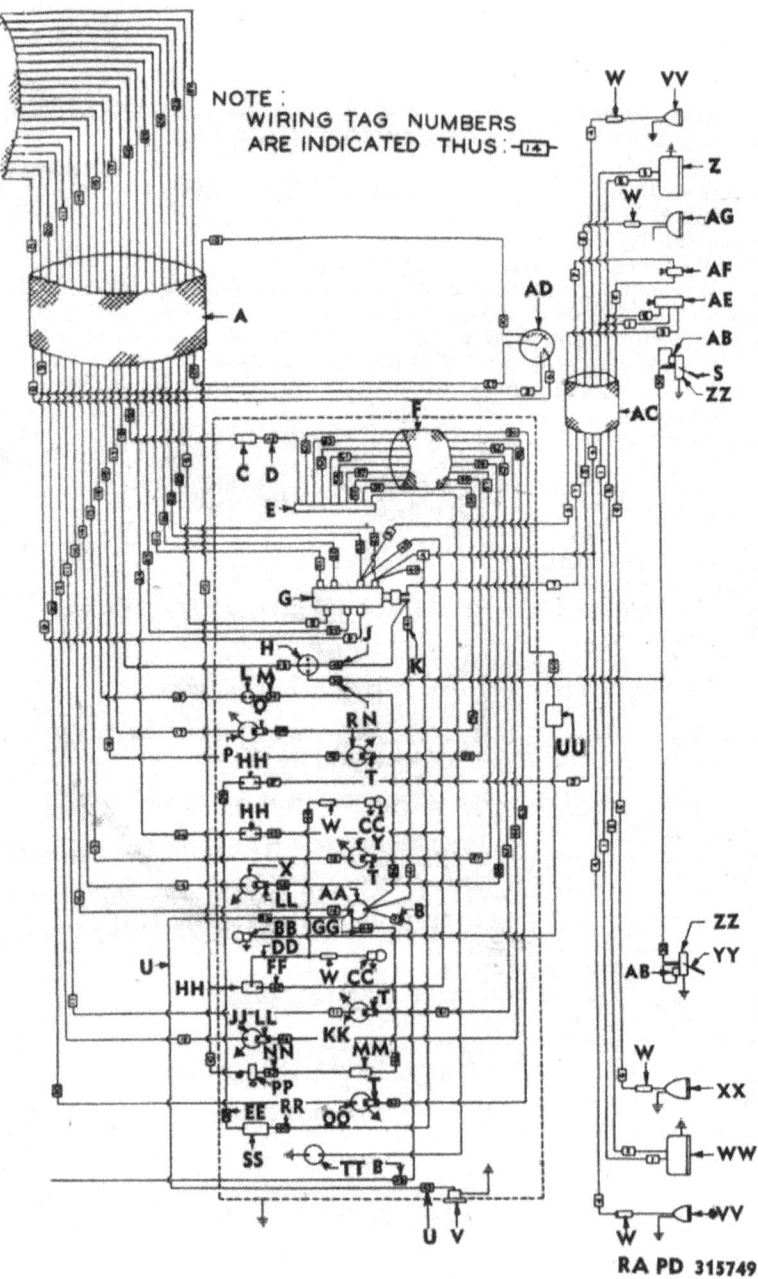

Figure 93—Wiring Diagram—Front Part of Tractor

ELECTRICAL SYSTEM

A—MAIN WIRING HARNESS (TAG NOS. 8 THRU 27 AND NO. 62)
B—RADIO FILTER TO AMMETER WIRE NEGATIVE (TAG NO. 29)
C—LIGHT FUSE (20 AMP.)
D—LIGHT FUSE WIRE (TAG NO. 42)
E—INSTRUMENT WIRE JUNCTION BLOCK
F—INSTRUMENT PANEL WIRING HARNESS (TAG NOS. 55 THRU 63)
G—MAIN SERVICE LIGHT SWITCH
H—WINDSHIELD WIPER SWITCH
J—WINDSHIELD WIPER SWITCH TO MAIN LIGHT SWITCH WIRE (TAG NO. 42)
K—AMMETER TO MAIN LIGHT SWITCH WIRE (TAG NO. 41)
L—CRANKING MOTOR PUSH-BUTTON SWITCH
M—AMMETER TO CRANKING MOTOR BUTTON SWITCH WIRE (TAG NO. 36)
N—ENGINE TEMPERATURE GAGE
P—ENGINE OIL PRESSURE GAGE
Q—ENGINE TEMPERATURE GAGE RESISTOR
R—ENGINE OIL PRESSURE GAGE
S—WINDSHIELD WIPER BLADE
T—ENGINE OIL PRESSURE RESISTOR, TRANSMISSION OIL PRESSURE RESISTOR, TORQUE CONVERTER FLUID PRESSURE RESISTOR, AND FUEL GAGE
U—TROUBLE LIGHT SOCKET TO AMMETER WIRE (TAG NO. 63)
V—TROUBLE LIGHT SOCKET
W—WIRING CONNECTOR (2-LINE)
X—TORQUE CONVERTER FLUID PRESSURE GAGE
Y—TORQUE CONVERTER FLUID PRESSURE GAGE
Z—HEADLIGHT, LEFT
AA—AMMETER ASSEMBLY
BB—LOW AIR PRESSURE INDICATOR LAMP
CC—DASH LAMP
DD—DASH LAMP WIRE (TAG NO. 44)
EE—BLACKOUT DRIVING LAMP SWITCH TO RESISTOR WIRE (TAG NO. 39)
FF—DASH SWITCH CONNECTING WIRE (TAG NO. 45)
GG—AMMETER TO ELECTRIC BRAKE RESISTOR WIRE (TAG NO. 46)
HH—BLACKOUT DRIVING LIGHT SWITCH, REAR FLOOD AND INSTRUMENT LAMP
JJ—TRANSMISSION AND OIL TEMPERATURE GAGE
KK—TRANSMISSION OIL PRESSURE GAGE
LL—TRANSMISSION OIL AND TORQUE CONVERTER FLUID PRESSURE GAGE RESISTOR
MM—ELECTRIC BRAKE RESISTANCE COIL
NN—ELECTRIC BRAKE RESISTOR TO LOAD CONTROL WIRE (TAG NO. 47)
PP—ELECTRIC BRAKE LOAD CONTROL
QQ—FUEL GAGE
RR—RESISTOR TO MAIN LIGHT SWITCH WIRE (TAG NO. 40)
SS—BLACKOUT DRIVING LIGHT RESISTOR
TT—HOUR RECORDING METER
UU—LOW AIR PRESSURE INDICATOR
VV—BLACKOUT MARKER LAMP
WW—HEADLAMP, RIGHT
XX—SIREN
YY—WINDSHIELD WIPER ARM
ZZ—WINDSHIELD WIPER
AB—WINDSHIELD WIPER RADIO SUPPRESSION CAPACITOR
AC—FRONT WIRING HARNESS (TAG NO. 1 THRU 7)
AD—ELECTRIC BRAKE CONTROLLER
AE—DIMMER SWITCH
AF—SIREN SWITCH
AG—BLACKOUT DRIVING LAMP

RA PD 315749 B

Legend for Figure 93—Wiring Diagram—Front Part of Tractor

18-TON HIGH SPEED TRACTOR M4

b. **Repair and Replacement of Wires.**

(1) REPAIR. A broken wire may be spliced by stripping about an inch of insulation off both broken ends of the wire and twisting the ends tightly together. Then wrap several thicknesses of friction or rubber tape around spliced section for insulation. If insulation is worn off or frayed from rubbing on metal, wrap the frayed or bare wire with tape.

(2) REPLACEMENT. Replacement of wires is seldom necessary unless the wires are destroyed by fire or like causes. When replacing any or all of the wires, remove the metal numbered tags from the wire or wires removed and install them on the corresponding new wires unless the new wire already has a tag with the same number on it. Be sure to tighten terminal connections firmly when installing wires.

Section XX

INSTRUMENTS AND GAGES

	Paragraph
General	98
Air pressure gage	99
Ammeter	100
Engine lubricating oil pressure gage	101
Engine temperature gage	102
Engine tachometer	103
Fuel gage	104
Hour meter	105
Low air pressure indicator	106
Speedometer	107
Torque converter fluid pressure gage	108
Torque converter fluid temperature gage	109
Transmission oil pressure gage	110
Transmission oil temperature gage	111

98. GENERAL.

a. The instruments are all mounted in the instrument panel on dash (fig. 96). These instruments and gages register pressures, temperatures, speeds, etc. of various operating parts of the tractor. By observing these instruments, the driver can readily know at all times whether or not the various units are functioning properly. When any of the gages register pressures, etc. not within the normal ranges as given in following paragraphs, the tractor must be stopped immediately and the cause determined and corrected.

99. AIR PRESSURE GAGE.

a. **Description.** The air pressure gage is a dial gage mounted on instrument panel and indicates the pressure in pounds per square inch of the air in air reservoir. Normal air pressure range is from 85 to 105 pounds while tractor is in operation.

b. **Replacement.**

(1) REMOVE GAGE. Open drain cock on air reservoir and release air from reservoir. Reach up behind instrument panel and disconnect air line from back of gage. NOTE: *Use care not to let wrench contact electrical terminals behind instrument panel.* Remove two nuts holding gage bracket and remove gage and bracket from panel.

(2) INSTALL REPLACEMENT GAGE. Use reverse of removal procedure to install replacement gage. After installation is completed,

TM 9-785
94

18-TON HIGH SPEED TRACTOR M4

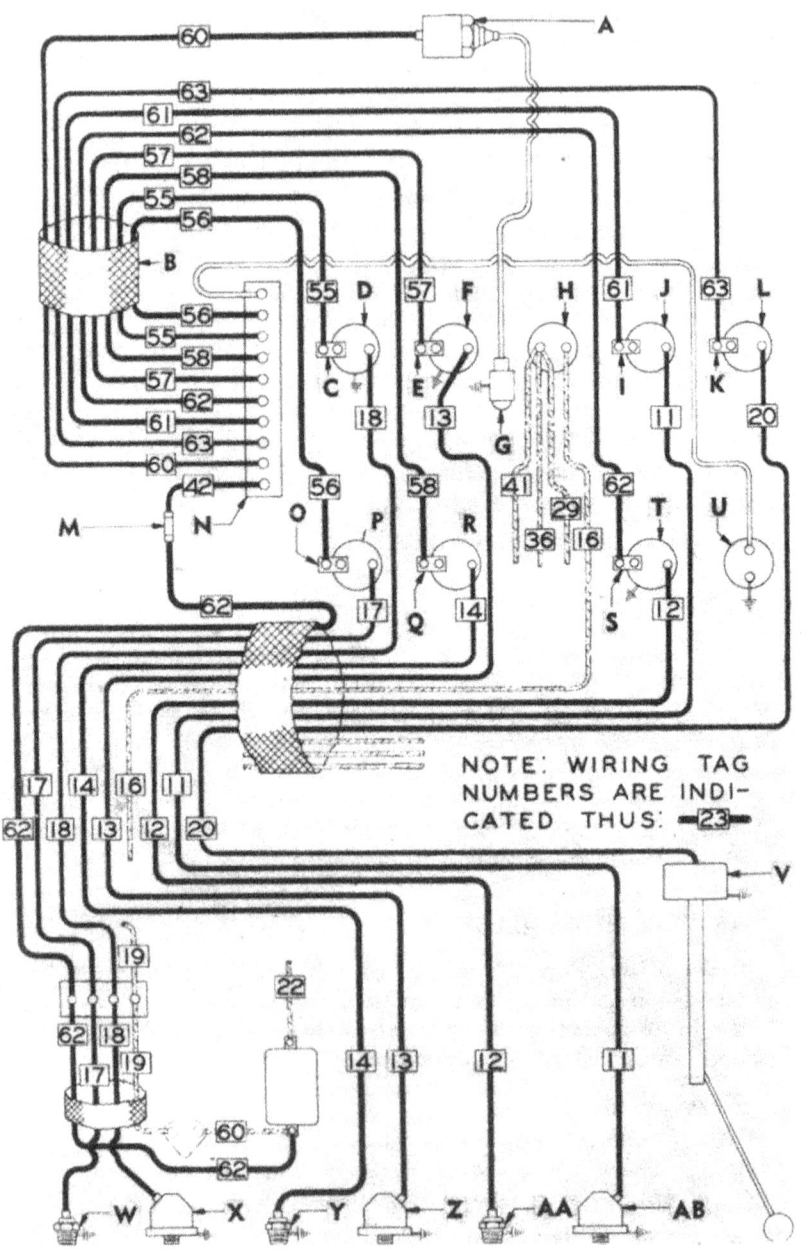

Figure 94—Wiring Diagram—Instruments

INSTRUMENTS AND GAGES

A—LOW AIR PRESSURE INDICATOR SWITCH
B—INSTRUMENT PANEL WIRING HARNESS
C—ENGINE OIL PRESSURE GAGE RESISTOR
D—ENGINE OIL PRESSURE GAGE
E—TORQUE CONVERTER FLUID PRESSURE GAGE RESISTOR
F—TORQUE CONVERTER FLUID PRESSURE GAGE
G—LOW AIR PRESSURE INDICATOR LAMP
H—AMMETER
I—TRANSMISSION OIL PRESSURE GAGE RESISTOR
J—TRANSMISSION OIL PRESSURE GAGE
K—FUEL GAGE RESISTOR
L—FUEL GAGE
M—LAMP FUSE
N—INSTRUMENT WIRE JUNCTION BLOCK
O—ENGINE TEMPERATURE GAGE RESISTOR
P—ENGINE TEMPERATURE GAGE
Q—TORQUE CONVERTER FLUID TEMPERATURE GAGE RESISTOR
R—TORQUE CONVERTER FLUID TEMPERATURE GAGE
S—TRANSMISSION OIL TEMPERATURE GAGE RESISTOR
T—TRANSMISSION OIL TEMPERATURE GAGE
U—HOUR RECORDING METER
V—FUEL TANK GAGE UNIT
W—ENGINE TEMPERATURE GAGE UNIT
X—ENGINE OIL PRESSURE GAGE UNIT
Y—TORQUE CONVERTER FLUID TEMPERATURE GAGE UNIT
Z—TORQUE CONVERTER FLUID PRESSURE GAGE UNIT
AA—TRANSMISSION OIL TEMPERATURE GAGE UNIT
AB—TRANSMISSION OIL PRESSURE GAGE UNIT

RA PD 307677B

Legend for Figure 94—Wiring Diagram—Instruments

close drain cock in reservoir, start engine to bring up pressure, and check for leakage at connection of air line.

100. AMMETER.

a. **Description.** The ammeter is a dial gage which registers the amount of current being delivered to the battery by the generator when vehicle is in operation (refer to par. 90).

b. **Replacement.**

(1) DISCONNECT INSTRUMENT PANEL FROM DASH. Remove four cap screws from wire guard on dash under panel (fig. 95). Disconnect air line from air pressure gage at rear of gage. Disconnect suction fuel line from engine primer pump. Disconnect discharge fuel line from engine primer pump. Remove four bolts holding instrument panel to dash and lay instrument panel back. Block up under panel (fig. 97) to take weight of panel off windshield wiper wire.

TM 9-785
100-101

18-TON HIGH SPEED TRACTOR M4

(2) REPLACE AMMETER. Remove four nuts and lift wires from terminals on ammeter. Tape end of wire leading to ammeter from battery. Remove two nuts holding ammeter mounting bracket and remove bracket and ammeter. To install replacement ammeter, reverse the removal procedure. Refer to wiring diagram (fig. 94) if necessary.

Figure 95—Removing Bolts from Wire Guard

101. ENGINE LUBRICATING OIL PRESSURE GAGE.

a. **Description.** The engine oil pressure gage is an electrically operated dial gage, mounted on instrument panel, with an operating unit mounted on engine. The gage hand registers the pressure of the lubricating oil delivered to the engine. Normal oil pressure is approximately 40 pounds with engine running at operating speed.

b. **Replacement of Gage.** Refer to paragraph 100 b. Use same procedure to replace the engine oil pressure gage as was used to replace ammeter.

c. **Replacement of Gage Operating Unit.**

(1) REMOVE OPERATING UNIT FROM ENGINE. Remove four cap screws holding right rear seat back cushion to rear of cab and remove cushion. Reach through opening in back of cab and disconnect wire from operating unit on side of engine cylinder block, below coil bracket. Unscrew unit from cylinder block.

196

INSTRUMENTS AND GAGES

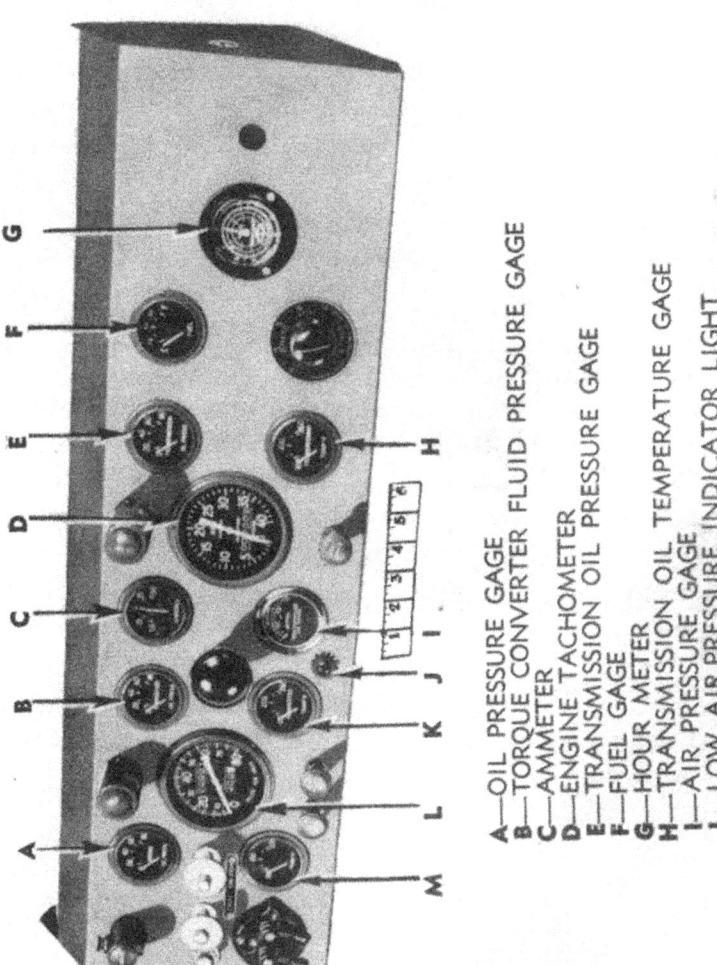

A—OIL PRESSURE GAGE
B—TORQUE CONVERTER FLUID PRESSURE GAGE
C—AMMETER
D—ENGINE TACHOMETER
E—TRANSMISSION OIL PRESSURE GAGE
F—FUEL GAGE
G—HOUR METER
H—TRANSMISSION OIL TEMPERATURE GAGE
I—AIR PRESSURE GAGE
J—LOW AIR PRESSURE INDICATOR LIGHT
K—TORQUE CONVERTER FLUID TEMPERATURE GAGE
L—SPEEDOMETER
M—ENGINE TEMPERATURE GAGE

Figure 96—Instrument Panel—Front View

(2) INSTALL NEW UNIT. Reverse procedure in (1) to install replacement unit. Use a sealing compound on threads of unit when installing it in engine cylinder block.

102. ENGINE TEMPERATURE GAGE.

a. **Description.** The engine temperature gage is an electrically operated unit, mounted on instrument panel and operated by a thermal unit mounted in top of engine water outlet manifold. The gage indicates the temperature of the engine coolant in degrees Fahrenheit. Normal engine temperature is from 160° to 180°F.

b. **Replacement of Gage.** Use same procedure as outlined in paragraph 100 b. "Replacement of ammeter," to replace temperature gage.

c. **Replacement of Gage Operating Unit.** Open engine grille and drain cooling system. Disconnect wire from unit after removing nut from terminal on unit. Unscrew unit from water outlet manifold. Install replacement unit by reversing removal procedure. **NOTE:** *Do not use any sealing compound on threads.*

103. ENGINE TACHOMETER.

a. **Description.** The engine tachometer, in instrument panel, registers the speed of the engine in hundreds of revolutions per minute of the engine crankshaft. It is actuated by a flexible drive shaft driven by a gear on the engine camshaft. The tachometer drive shaft is enclosed in a flexible housing.

b. **Replacement of Tachometer.** Use same procedure for replacing tachometer as outlined in paragraph 107 b. "Replacement of speedometer."

c. **Replacement of Tachometer Drive Shaft.** Reach behind instrument panel and disconnect tachometer drive shaft and housing from tachometer. Remove four cap screws holding right rear seat back cushion to back of cab and remove cushion. Reach through opening in back of cab and disconnect drive shaft and housing from tachometer drive housing on engine. Grasp rear end of shaft and pull end out of flexible housing far enough to remove spring clip from shaft and remove clip. Grasp front end of shaft and pull shaft from housing. Reverse removal procedure to install new drive assembly.

104. FUEL GAGE.

a. **Description.** The fuel gage in instrument panel is an electrically operated gage with an operating unit mounted in top of fuel tank under rear floor plate. This gage registers the level of the fuel in the tank. It will operate only when ignition switch is closed.

b. **Replacement of Gage** Use same procedure to replace fuel gage as used to replace ammeter, paragraph 100 b.

INSTRUMENTS AND GAGES

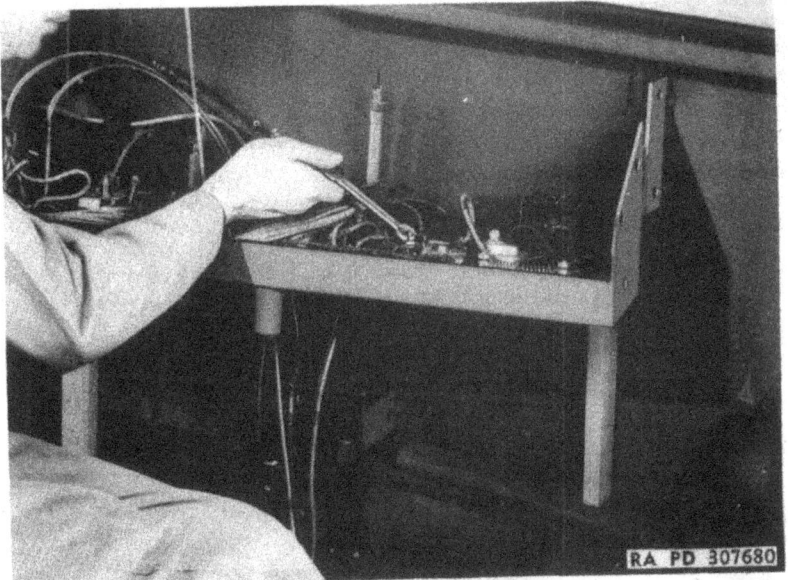

Figure 97—Disconnecting Wires from Fuel Gage

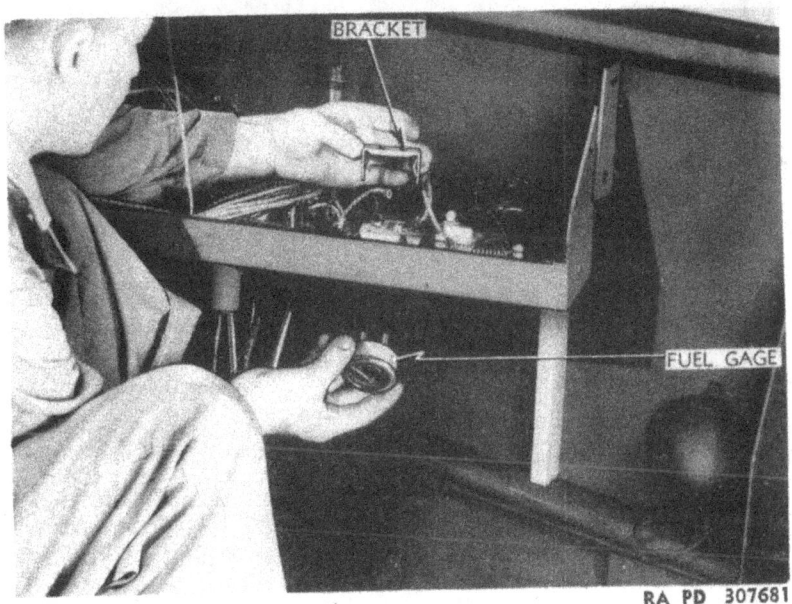

Figure 98—Fuel Gage and Bracket Removed

199

TM 9-785
104-105

18-TON HIGH SPEED TRACTOR M4

c. **Replacement of Operating Unit.** Remove seat cushions and seat plates from rear compartment. Raise rear floor plate and fasten with snap fastener. Remove nut and lift wire from terminal of unit. Remove six screws holding unit to tank. Remove unit from tank by lifting up and out, maneuvering unit carefully so float and rod will not be damaged in removal. Reverse removal procedure to install replacement unit.

105. HOUR METER.

a. **Description.** The hour meter is an electrically operated meter which registers the total number of hours engine has operated (par. 17).

Figure 99—Removing Hour Meter Retaining Screws

b. **Replacement of Hour Meter.**

(1) DISCONNECT INSTRUMENT PANEL FROM DASH. Refer to procedure under paragraph 100 b.

(2) REMOVE HOUR METER. Remove three cross-recessed screws holding meter in instrument panel (fig. 99). Remove long bolt holding resistance unit to brackets to which ground wire of hour meter is fastened. Lay resistance unit away from brackets. Remove bolt holding ground wire and bracket to instrument panel. Pull end of remaining wire attached to hour meter from rear socket of junction

INSTRUMENTS AND GAGES

block at left end of panel, then out through clips. Remove hour meter.

(3) INSTALL REPLACEMENT METER. Reverse removal procedure to install replacement meter. Refer to wiring diagram (fig. 94) for proper wire connections if necessary.

106. LOW AIR PRESSURE INDICATOR.

a. **Description.** This indicator consists of a small light with a red lens, mounted in center of instrument panel. It is operated by a pressure switch in the air line. It lights up when ignition switch is closed to warn the operator when the air pressure in the air

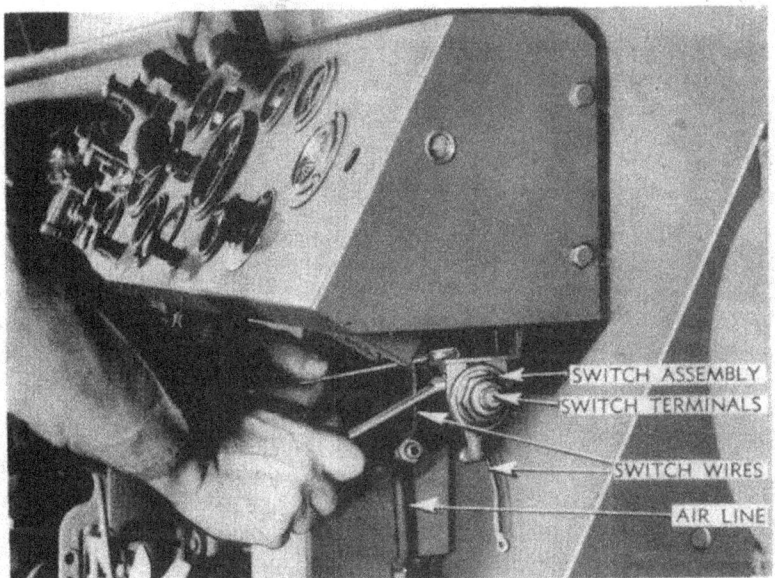

Figure 100—Removing Bolts from Switch

reservoir is too low for the air brakes on the trailed vehicle to operate efficiently and remains lighted until the pressure raises to approximately 65 pounds. The automatic switch then opens and light remains off until pressure again drops to about 55 pounds.

b. **Replacement of Bulb.** Reach behind instrument panel and pull bulb socket from light. Turn bulb to left ⅛ turn and pull bulb from socket. Insert new bulb in socket in reverse order.

c. **Replacement of Low Air Pressure Indicator Switch.** Disconnect air line from switch. Remove two nuts and disconnect the two wires from switch. Remove two bolts holding switch bracket on

TM 9-785
106-107

18-TON HIGH SPEED TRACTOR M4

dash (fig. 100) and remove switch. Reverse procedure to install new switch.

107. SPEEDOMETER.

a. **Description.** The needle of the speedometer in instrument panel registers speed of travel in miles per hour, the upper figures register total miles vehicle has travelled, and the lower figures can be set to register miles travelled for each trip, if desired. The speedometer is operated by a flexible drive shaft from the bevel pinion shaft of the transmission.

b. **Replacement of Speedometer.** Disconnect instrument panel as outlined in paragraph 100 b. Disconnect drive shaft housing from

Figure 101—Disconnecting Cable from Speedometer

speedometer (fig. 101). Remove two wing nuts from mounting bracket and remove bracket and speedometer (fig. 102). Install replacement unit, reversing removal procedure.

c. **Replacement of Speedometer Drive Shaft.** Remove seat cushions and seat plates from rear compartment. Raise floor plate and fasten with snap fastener. Disconnect one end of drive shaft housing from drive housing at back of transmission and the other end from rear of speedometer. Instrument panel is laid back in pictures only to show connections; it is not necessary for removal or installation.

INSTRUMENTS AND GAGES

Grasp lower end of shaft and pull end out of flexible housing far enough to remove spring clip from shaft and remove clip. Pull shaft out of housing from upper end. Install new flexible drive shaft assembly by reversing procedure.

108. **TORQUE CONVERTER FLUID PRESSURE GAGE.**

a. Description. The torque converter fluid pressure gage is an electrically operated gage mounted on instrument panel and registers the operating pressure of the fluid in the converter system. It is actuated by an operating unit on torque converter housing.

b. Replacement of Gage. Use same procedure to replace this gage as used to remove ammeter, paragraph 100 b.

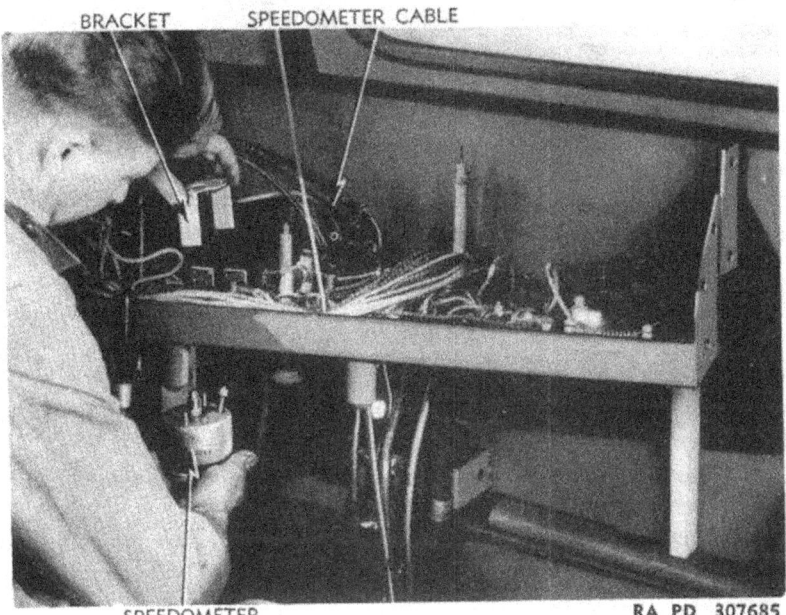

Figure 102—Removing Speedometer

c. Replace Gage Operating Unit. Remove rear compartment seat cushions and seat plates. Raise rear floor plate and fasten with snap fastener. Remove cap screws from rear compartment front floor plate and remove plate. Remove nut and lift wire from terminal of unit (fig. 106). Have new unit close at hand. Unscrew old unit from torque converter until it is loose enough to turn with fingers. Then quickly remove it and start new one into torque converter in its place. Tighten with wrench and connect wire to terminal on unit. Install front floor plate, lower hinged floor plate, and place seat plates and cushions back in tractor.

18-TON HIGH SPEED TRACTOR M4

109. TORQUE CONVERTER FLUID TEMPERATURE GAGE.

a. **Description.** The torque converter fluid temperature gage in instrument panel is an electrically operated gage that registers the temperature of the fluid in the converter system in degrees Fahrenheit. A thermal unit that operates gage is mounted at lower left side of converter in fluid manifold.

b. **Replacement of Gage.** Use the same procedure to replace the gage as is used to replace ammeter, paragraph 100 b.

c. **Replacement of Thermal Unit.** Follow the same procedure as outlined under paragraph 105 c to replace thermal unit.

110. TRANSMISSION OIL PRESSURE GAGE.

a. **Description.** The transmission oil pressure gage is an electrically operated unit that registers pressure of transmission oil. It is actuated by an operating unit mounted in transmission oil distribution manifold under front floor plate of rear seat compartment.

b. **Replacement of Gage.** Use same procedure as for replacing ammeter (par. 100 b).

c. **Replacement of Gage Operating Unit.** Use same procedure as outlined under paragraph 108 c to replace operating unit.

111. TRANSMISSION OIL TEMPERATURE GAGE.

a. **Description.** The transmission oil temperature gage in instrument panel is electrically operated by a thermal unit. This gage registers the temperature of the transmission oil in degrees Fahrenheit. The thermal unit is located in oil distribution manifold under front floor plate of rear seat compartment (fig. 115).

b. **Replacement of Gage.** Use same procedure to replace gage as used to replace ammeter, paragraph 100 b.

c. **Replacement of Thermal Unit.** Follow same procedure as outlined under paragraph 105 c to replace thermal unit.

TM 9-785
112-113

Section XXI

MASTER CLUTCH

	Paragraph
Description	112
Clutch pedal adjustment	113
Replacement of clutch drive disk	114

112. DESCRIPTION.

a. The master clutch is of the spring-loaded, dry-disk type. A driven disk, with friction lining on both sides, is riveted to a splined hub carried on the master clutch shaft. All other parts of the clutch assembly except the release mechanism are bolted to the engine flywheel. The clutch is held engaged by springs between pressure plate and back plate and disengaged by forcing pressure plate away from the drive disk by depressing the clutch pedal. This actuated the release mechanism in clutch housing and relieves the pressure of the flywheel and pressure plate against the drive disk. A flexible grease tube extends from outside the clutch housing to the release bearing for lubrication of the bearing and sleeve.

113. CLUTCH PEDAL ADJUSTMENT.

a. General. The master clutch pedal must have 1½-inch "free travel" at all times to insure clearance between the release bearing and throwout fingers in the back plate assembly and to insure complete engagement and disengagement of the clutch. The adjustment for this free travel is made by shortening or lengthening the control rod between clutch pedal lever bell crank and clutch release yoke shaft in clutch housing (fig. 116). No other adjustments will be made by operating organization.

b. Adjustment of Free Pedal Travel.

(1) MEASURE FREE TRAVEL OF PEDAL. With clutch pedal against stop, place one end of ruler against dash and other end over top of clutch pedal. Observe measurement. Depress clutch pedal until contact of release bearing against throwout fingers is felt. Note measurement. If difference between the two measurements is less than 1½ inch, control rod (fig. 116) must be lengthened. If free travel is more than 1½ inch, it is possible that clutch will not be completely disengaged when pedal is depressed and control rod must be shortened.

(2) ADJUST CONTROL ROD. Remove seat cushions and seat plates from rear seat compartment. Lift rear floor plate and fasten open with snap fastener. Remove five cap screws and one bolt from rear compartment front floor plate and remove plate. Loosen jam nut on rod at yoke connecting front end of control rod to bell crank on power take-off housing (fig. 116). Pull cotter pin from yoke pin connecting control rod to lever on clutch release yoke shaft and re-

TM 9-785
113-114

18-TON HIGH SPEED TRACTOR M4

move yoke pin. Turn rod out of yoke to lengthen rod and increase free travel of pedal, or into yoke to shorten rod and decrease free travel, whichever is required. Connect rod to lever on clutch release yoke shaft and check as in (1). When desired free travel is obtained, install cotter pin in control rod yoke pin and tighten jam nut on rod against yoke. Install front floor plate, lower rear floor plate, and install seat plates and cushions.

Figure 103—Master Clutch Assembly

114. REPLACEMENT OF CLUTCH DRIVE DISK.

a. **Remove Engine from Tractor.** Follow outlined procedure in paragraph 62 for removal of engine.

b. **Remove Clutch Assembly from Flywheel** (fig. 38). Install six $\frac{3}{8}$- x $1\frac{1}{2}$-inch NC cap screws through holes in clutch cover plate and screw them into tapped holes in release lever bosses of pressure plate until the heads of the cap screws are against cover plate. These are to hold clutch springs compressed while clutch assembly is

MASTER CLUTCH

removed. Then remove the eighteen cap screws attaching clutch cover plate to flywheel and remove cover plate assembly. Clutch drive disk will be removed at same time. Be careful not to lose the three dowels in flywheel which extend through cover plate and may pull out of flywheel when cover plate is removed. Inspect all parts in cover plate and pressure plate assembly and in clutch housing while clutch and engine is removed to determine if any of those parts are worn or damaged and in need of replacement.

c. **Install Replacement Clutch Assembly.** Insert the three small dowels with snap rings into holes in flywheel, seating the snap rings in recesses in flywheel. Set drive disk against flywheel with shorter end of hub towards flywheel, then place cover plate in position on dowels. Start the eighteen cap screws with lock washers that attach cover plate to flywheel. Aline hub of drive disk with clutch shaft pilot bearing and tighten cap screws.

d. **Install Engine in Tractor.** Follow installation instructions in paragraph 63 to install engine.

TM 9-785
115

18-TON HIGH SPEED TRACTOR M4

Section XXII

TORQUE CONVERTER AND PROPELLER SHAFT

	Paragraph
Propeller shaft	115
Torque converter	116
Torque converter fluid filter	117
Torque converter fluid cooling radiator	118

115. PROPELLER SHAFT.

a. Description. The propeller shaft (fig. 104) consists of two universal joint assemblies with attaching flanges and a short shaft.

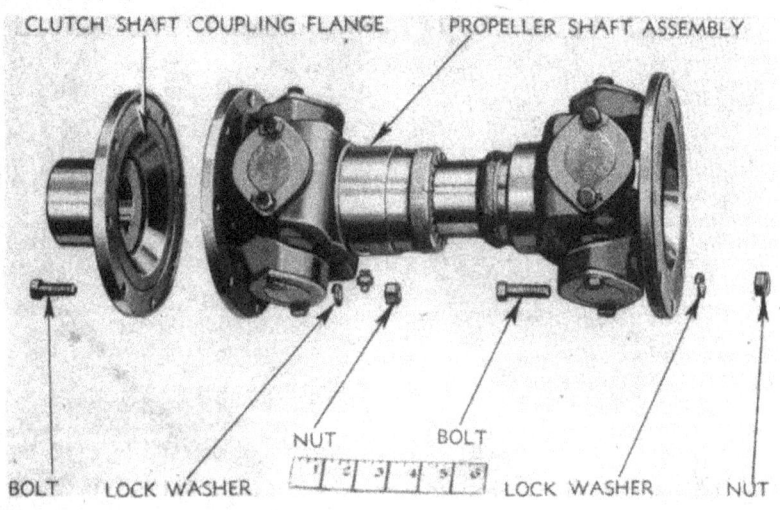

Figure 104—Propeller Shaft Assembly

The attaching flange at front end of propeller shaft is bolted to a flange on pump shaft of torque converter, the flange at rear end is bolted to a coupling flange on clutch shaft. The universals are equipped with needle roller bearings.

b. Replacement of Propeller Shaft Assembly. Remove seats from rear seat compartment. Lift hinged floor plate and fasten open with snap fastener. Remove five cap screws and one bolt and remove

TM 9-785
115-116

TORQUE CONVERTER AND PROPELLER SHAFT

floor plate above torque converter. Remove nuts and lock washers from bolts attaching flange on front universal to flange of torque converter pump shaft (fig. 105). Remove the eight bolts attaching flange of rear universal joint to coupling flange on clutch shaft and remove propeller shaft assembly from tractor (fig. 106). Install replacement unit by reversing removal operations. Tighten bolts firmly.

116. TORQUE CONVERTER.

a. Description.

(1) PURPOSE AND LOCATION. Torque can be defined briefly as the effort producing, or attempting to produce, rotation or torsion, and is the product of force exerted to move an object times distance the object is moved. When torque is produced at a given rate of speed, horsepower is developed and, within an elapsed time, work is

Figure 105—Disconnecting Propeller Shaft from Torque Converter

accomplished. The relationship of horsepower to torque and speed is such that for a given amount of horsepower an increase in torque will be compensated by a decrease in speed. Since an internal combustion engine is unable to produce torque effectively until it is operating at a certain minimum speed, it is necessary to convert the torque developed by the engine in order to apply it in starting a heavy

209

18-TON HIGH SPEED TRACTOR M4

load. The hydraulic torque converter, therefore, is used to develop the required amount of torque to start a heavy load which the engine is capable of pulling after tractor is in motion. The torque converter also allows smooth and shockless starting of load and acceleration, prevents slipping of clutch in starting load, and prevents stalling of engine under heavy pulls. It is mounted at rear of transmission, the input shaft is connected to propeller shaft, and output shaft is connected to transmission input shaft.

(2) PRINCIPLES OF OPERATION. The Twin-Disc Model T-10010 torque converter consists basically of three major components; the pump or impeller, the turbine, and stationary housing. The fluid which completely fills the converter is the medium for transferring

Figure 106—Removing Propeller Shaft Assembly

the power of the engine to the driving sprockets through these parts. The operation of the torque converter depends upon the circulatory movement of the fluid for the transmission of power. This movement is accomplished in the following manner. The converter pump or impeller, which is coupled to the engine by the propeller shaft and clutch shaft, circulates the fluid, and the fluid, through the velocity imparted to it by the pump, becomes a transmission medium for the power delivered by the engine. As the fluid leaves the impeller it has attained its maximum velocity and immediately encounters the first set of blades of the turbine. Upon contacting this first set of tur-

TORQUE CONVERTER AND PROPELLER SHAFT

bine blades, the fluid releases a portion of its energy imparted to it by the impeller and starts the turbine to rotate. It must be remembered that not all of the energy imparted to the fluid by the pump is released against the first set of turbine blades; therefore, in order to place the fluid in a position to release the remaining energy, reactionary or guide blades are riveted in the stationary housing. The reactionary blades redirect the flow of fluid so that it enters the second set of turbine blades at the correct angle. As the fluid passes through the second set of turbine blades, additional energy is released imparting more torque on the turbine, whereupon the fluid continues its circulation toward the third and last set of turbine blades. However, before the fluid can enter the third set of turbine blades, it must pass through a second set of reactionary blades for the same reason given above. As the fluid leaves the third set of turbine blades, it has released very nearly all of its energy and it is exhausted from the outlet of the turbine directly into the inlet of the pump or impeller wherein the same cycle of circulation is repeated.

(3) COOLING SYSTEM. As various load conditions are encountered by the tractor, the temperature of the fluid can be expected to rise to the point where it becomes necessary for it to be cooled. A separate radiator has been provided in radiator assembly (outer radiator) for cooling the converter fluid. The fluid is discharged from converter into a line that extends from the outside diameter of the converter to the radiator. After circulating through radiator, the fluid returns to the center of the converter. As previously mentioned, the fluid velocity is greatest as the fluid leaves the impeller, and is the least as the fluid enters the impeller near the center of the converter. This difference in pressure is utilized to circulate the fluid through the radiator.

(4) RESERVE TANK. Since the torque converter must be completely filled with fluid for satisfactory operation, a reserve tank is provided to take care of the volumetric increase of the fluid as the temperature rises. This tank is provided with an orifice connected to a bleed line extending from the top of the fluid radiator which remains open to the reserve tank at all times through which the expanded fluid escapes from the main fluid circuit. This orifice also serves as a means of relieving the converter and radiator of any gas or air which may become trapped within the converter or radiator. A reserve of approximately two gallons of fluid is carried in this tank to provide against loss of fluid due to vaporization or small leaks. The reserve tank is open to the atmosphere through a separate vent line.

(5) AUXILIARY FLUID PUMP. The purpose of the auxiliary fluid pump is twofold. First, the auxiliary pump returns that fluid to the converter which escapes into the reserve tank through the bleed line from the radiator. Secondly, the auxiliary pump prevents cavita-

18-TON HIGH SPEED TRACTOR M4

tion of the fluid at the inlet to the impeller which otherwise would prove detrimental to the efficient operation of the converter. A relief valve is used in conjunction with the auxiliary pump set at a pressure of approximately 40 to 50 pounds per square inch. The capacity of this pump is approximately 3.5 gallons per minute. A filter (fig. 108) is located in the suction line of the auxiliary pump from the reserve tank.

(6) FREEWHEEL ASSEMBLY. A freewheel assembly, consisting of an inner race, outer race, and cage assembly is incorporated in the converter assembly. The freewheel outer race is assembled to the turbine and actually forms a part of the hydraulic chamber. The freewheel cage assembly is composed of alternately spaced rollers and springs held within a cage. The freewheel inner race is mounted on the input shaft. The operation of the freewheel group assembly is unique in that it serves both as a pilot bearing and as a freewheel unit. When the engine is under load pulling the tractor over rough terrain or along a highway, the freewheel group assembly functions as a bearing. If the tractor should encounter a downgrade and begin to overrun the engine, the freewheel group assembly locks and the impeller and turbine turn together, driving back against the engine. Thus, better braking is obtained, because in addition to using the engine as a brake, the drag of the converter also augments the braking action. The use of the freewheel group assembly again comes into the picture when trying to start the engine by towing or pushing the tractor. Normally, with a fluid drive, such a process is impracticable as it requires towing or pushing the vehicle at too great a speed; however, starting the engine by towing or pushing is possible with the freewheel group assembly as the freewheel locks and the drive is taken back to the engine.

b. **Maintenance.**

(1) Under normal operating conditions, due to the atmosphere and to heat, the fluid will eventually show signs of oxidation and carbon formation. NOTE: *Spare torque converter fluid should always be kept in clean, airtight container in order to assure protection from water and foreign material.*

(2) FILL SYSTEM. Make sure drain plug in bottom of converter housing is securely in place, drain valve at bottom of fluid reserve is closed, fitting at rear of filter in bleeder line (fig. 110) is tight. Remove reserve tank filler cap and bayonet gage. Fill tank with fluid (either kerosene with minimum initial boiling point of 350°F, or approximately No. 2 Diesel fuel oil). Start engine, then continue to pour fluid into tank as system fills until 35½ U.S. quarts of fluid have been used. Pointer of converter fluid pressure gage will flutter and not give a steady reading until the circulation of fluid has removed all air pockets from system. Fluid level now should be even with "FULL" mark on reserve tank bayonet gage. Add or drain fluid

TORQUE CONVERTER AND PROPELLER SHAFT

to bring level to "FULL" mark. Inspect for possible leaks in fluid line connections after filling system.

(3) DRAIN SYSTEM. Remove seats from rear seat compartment and raise hinged floor plate. Remove drain plugs from bottom of hull under torque converter assembly. Remove drain plug from bottom of torque converter and open drain valve at bottom of fluid reserve tank. Loosen pipe fitting at rear of filter in bleeder line (fig. 110) to allow air to enter system. Tighten this fitting as soon as system is drained.

c. **Replacement of Torque Converter.**

(1) REMOVE PROPELLER SHAFT ASSEMBLY. Follow procedure outlined in paragraph 115 b to remove propeller shaft assembly.

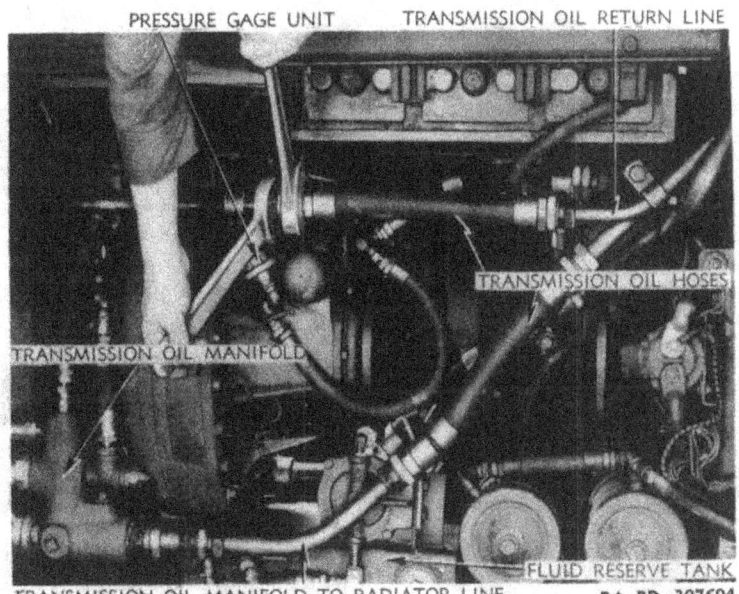

Figure 107—Disconnecting Hose in Transmission Oil Return Line

(2) DRAIN CONVERTER SYSTEM. Refer to b (3) for procedure.

(3) DISCONNECT BATTERY AND WIRES. Disconnect one of the battery cables from battery. Disconnect wire from terminal of converter fluid pressure gage unit at top of converter and wire from terminal of temperature gage unit between reserve tank and converter.

(4) REMOVE OR DISCONNECT UPPER HOSE AND LINES (fig. 107). Remove section of hose from transmission oil return line. Remove bolt from bracket and clips supporting converter fluid and trans-

18-TON HIGH SPEED TRACTOR M4

Figure 108—Disconnecting Torque Converter Fluid Filter Line

Figure 109—Disconnecting Converter Fluid Hose from Relief Valve

TORQUE CONVERTER AND PROPELLER SHAFT

mission oil hose. Remove section of hose and front section of pipe from transmission oil manifold to radiator line.

(5) *REMOVE CONVERTER FLUID FILTER (fig. 108). Disconnect both hose from filter connections. Remove the two mounting bolts from filter bracket and filter and remove filter assembly.

(6) DISCONNECT CONVERTER FLUID LINES FROM CONVERTER. Remove bolt attaching lower end of floor plate support angle at front corner of battery (fig. 108) to cross member on bottom of hull and remove angle. Disconnect rear end of hose connecting converter auxiliary fluid pump to relief valve at side of battery (fig. 109). Re-

Figure 110—Disconnecting Converter Fluid Hose from Pipes

move bolts from clips on the two large converter fluid hose. Disconnect rear ends of large hose from ends of pipes (fig. 110). Disconnect front bearing lubricating oil hose from fitting at front and top of converter (fig. 111). Disconnect hose between bottom of front end of reserve tank and T-fitting at front and bottom of converter from T-fitting.

(7) REMOVE CONVERTER FROM TRACTOR. Remove the twelve nuts and lock washers from studs attaching converter to transmission. Screw a ½-inch standard thread eyebolt into tapped hole in top of converter (fig. 111). Lower hoist hook through gun ring on top of cab, hook into eyebolt, slide converter from splines, and lift converter assembly from tractor (fig. 112).

215

18-TON HIGH SPEED TRACTOR M4

Figure 111—Disconnecting Lubricating Oil Hose from Converter

Figure 112—Lifting Converter Out of Tractor

TORQUE CONVERTER AND PROPELLER SHAFT

(8) INSTALL REPLACEMENT UNIT. Follow removal procedure in reverse order for installation of converter as follows: Lower converter into place, and attach to transmission (fig. 112). Connect hose from reserve tank to T-fitting at front bottom of converter. Connect lubricating oil hose to fitting at top front of converter (fig. 111). Connect rear ends of large converter fluid hose to pipes leading to radiator (fig. 110). Install two bolts with lock washers in hose clips on these hose and bolt clips to support (fig. 109). Connect hose from auxiliary pump to relief valve (fig. 109). Install floor plate support angle with one bolt with lock washer in lower end (fig. 108). Install fluid filter with two bolts and lock washers and connect fluid lines to filter (fig. 108). Connect transmission oil lines (fig. 107). Connect wires to terminals on fluid pressure and temperature gage

RA PD 307700

Figure 113—Disconnecting Fluid Lines from Auxiliary Pump

operating units, install propeller shaft (par. 115 b) and connect battery cable to battery terminal. Fill torque converter system (b (2)), then test for leaks in system, proper fluid pressure, and proper operation of converter.

d. **Replacement of Auxiliary Fluid Pump** (fig. 113). Disconnect hose from fittings on pump. Remove the four nuts and lock washers from attaching studs and remove pump from studs. Install new pump, using new gasket, and attach hose as shown.

TM 9-785
117-118

18-TON HIGH SPEED TRACTOR M4

117. TORQUE CONVERTER FLUID FILTER.

a. **Description.** The Commercial Model AS 4¼ DDV is located on a bracket at left and back of converter. It is of the replaceable element type. Its purpose is to remove foreign material from the fluid delivered to the converter system by the auxiliary pump on converter.

b. **Maintenance.** Clogging will be indicated by converter fluid pressure gage showing a drop from normal pressure. Drain screw in bottom must be loosened weekly and water or sediment allowed to drain from filter.

Figure 114—Installing Converter Fluid Filter Element

c. **Replacement of Fluid Filter Element** (fig. 114). Hold small pan under filter, loosen drain screw, and drain filter. Remove four cap screws clamping filter bowl to filter head and lift filter bowl and element from filter head. Remove element, wash bowl thoroughly and install new element. Use new gasket when installing bowl on filter head. Start engine after installation and check for leaks.

118. TORQUE CONVERTER FLUID COOLING RADIATOR.

a. The converter fluid cooling radiator is the outside radiator in the radiator assembly. It is of the fin and tube type similar to

TORQUE CONVERTER AND PROPELLER SHAFT

the water radiator. The fluid in converter system is circulated through the connecting lines and radiator by the torque converter as explained in paragraph 116 a (3) and is cooled as it passes through radiator by the air drawn through radiator by the fan.

b. **Replacement of Converter Radiator.** Follow procedure outlined in paragraph 83 c, d, e, and f to replace converter radiator.

18-TON HIGH SPEED TRACTOR M4

Section XXIII

TRANSMISSION, DIFFERENTIAL, AND FINAL DRIVES

	Paragraph
Description	119
Replacement of transmission oil pump	120
Replacement of transmission oil cooling radiator	121

119. DESCRIPTION.

a. Power supplied by the engine is transferred through the torque converter to the transmission, controlled differential, and final drives to the drive sprockets. The transmission provides for three forward speed ranges, an engine brake gear position for descending hills, and one reverse speed range (fig. 6). The power take-off assembly is also driven by the transmission gears. The transmission and differential are enclosed in separate cases, the final drives are bolted to the differential case, the power take-off to the transmission case. The gear arrangement in the differential not only drives the sprockets, but provides for steering tractor as well. Each side of the differential is controlled by a steering brake drum and brake band, which, when band is contracted around drum, causes tractor to turn. The final drive pinion shafts, the ends of which are inserted into differential hub, drive the larger final drive gears on sprocket shafts.

b. Lubrication. The final drives are lubricated by the oil carried in their respective housings. The transmission, differential, and power take-off assemblies are lubricated by a common oil supply, distributed by a pump mounted on power take-off case and driven by the power take-off gears. A filler plug and bayonet level gage are provided in differential cover (fig. 121) for filling to proper level. The oil pump draws the oil from a sump in bottom of differential case, circulates it through the oil cooling radiator where it is cooled and delivers it through a manifold and system of pipes to the various gears and bearings. A restricted fitting in the oil discharge line maintains pressure in lubricating system. The oil level in final drives must be maintained at a level even with filler plugs, the oil level in differential must be maintained between "LOW" and "FULL" marks on bayonet gage in differential cover. The using troops will maintain oil levels in final drives, transmission and differential, and will change lubricant at specified intervals. CAUTION: *Do not fill any oil compartments above level plugs or above "FULL" mark on bayonet.*

TRANSMISSION, DIFFERENTIAL, AND FINAL DRIVES

Figure 115—Disconnecting Hose from Oil Distribution Manifold

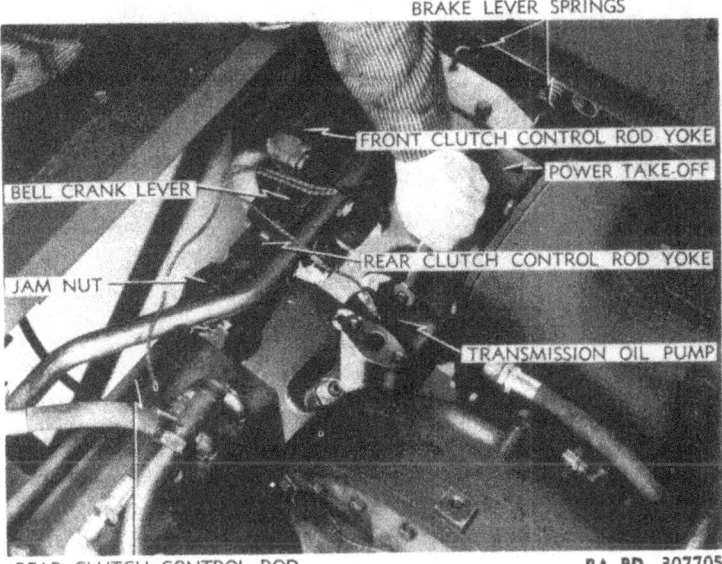

Figure 116—Removing Pump Mounting Cap Screws

TM 9-785
120

18-TON HIGH SPEED TRACTOR M4

120. REPLACEMENT OF TRANSMISSION OIL PUMP.

a. **Remove Oil Pump.** Remove seats from rear seat compartment and lift hinged floor plate. Remove cap screws holding floor plate section at front of rear seat compartment and remove plate. Remove nuts and washers and remove ends of wires from terminals of transmission oil temperature and pressure gage operating units on oil distribution manifold. Disconnect all hose connected to distribution manifold (fig. 115) and disconnect hose from fitting on lower left side of transmission case. Remove cap screw holding clip on above hose to power take-off case. Remove two cap screws holding distribution manifold to oil pump body and lift off manifold and oil lines. Remove two cap screws holding suction oil line to bottom of pump body (fig. 117). Remove four cap screws holding pump to power take-off case (fig. 116) and remove pump (fig. 117).

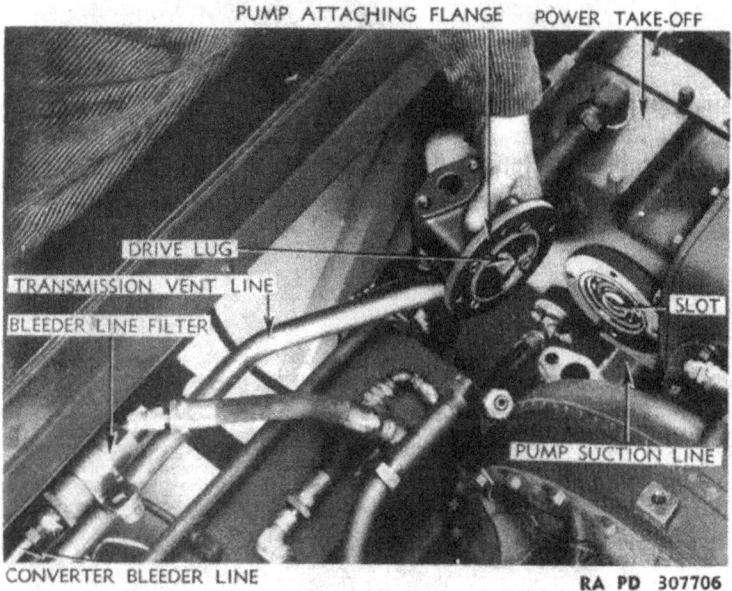

Figure 117—Oil Pump Removed

b. **Install Pump.** Shellac gasket to attaching flange of pump. Place pump in position on power take-off case with drive lug on pump shaft engaging in notch in end of power take-off shaft and smaller of the two openings in pump body (fig. 117). Attach pump to power take-off case with four cap screws with lock washers (fig. 116). Use new gasket between mating surfaces and connect suction oil line to

222

TRANSMISSION, DIFFERENTIAL, AND FINAL DRIVES

bottom of pump housing. Install distribution manifold on upper side of pump housing, using new gasket, with two cap screws and lock washers. Connect hose leading from manifold to lower left side of transmission case and install cap screw in clip on this hose and into power take-off case. Connect remaining hose to manifold as shown in figure 115. Connect wire from transmission oil temperature gage to terminal on temperature gage operating unit and wire from transmission oil pressure gage to terminal on pressure gage operating unit on manifold with nuts and washers. Start and operate engine to test operation of pump and check for leaks, then install front floor plate section and seats.

121. REPLACEMENT OF TRANSMISSION OIL COOLING RADIATOR.

a. It is necessary to remove entire radiator assembly to replace transmission oil radiator. Follow procedure outlined in paragraph 83 c, d, e, and f for replacement instructions.

18-TON HIGH SPEED TRACTOR M4

Section XXIV

STEERING BRAKES

	Paragraph
Description	122
Brake adjustment	123
Replacement of steering brakes	124

122. DESCRIPTION.

a. The steering brakes consist of two lined brake shoe assemblies which encircle the brake drums on differential end cover hubs and are operated by the two hand levers. Contracting the brake shoes of

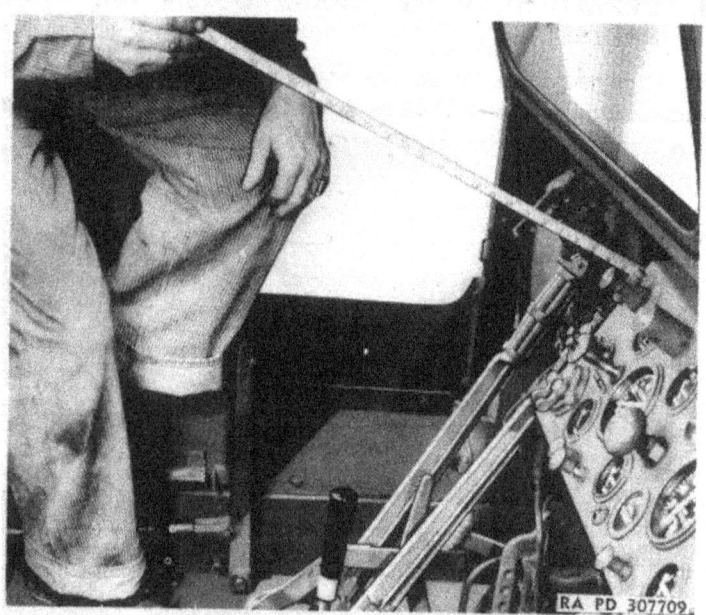

Figure 118—Measuring with Lever Forward

either brake around its drum sets the planetary gears in motion causing the track on that side to slow down and the track on other side to speed up and tractor is turned. Locks are provided on tops of levers to engage ratchets to lock brakes in contracted position for parking tractor.

123. BRAKE ADJUSTMENT.

a. The steering brakes are correctly adjusted and have the proper clearance when the top of the levers have 6 inches of free travel

TM 9-785
123-124

STEERING BRAKES

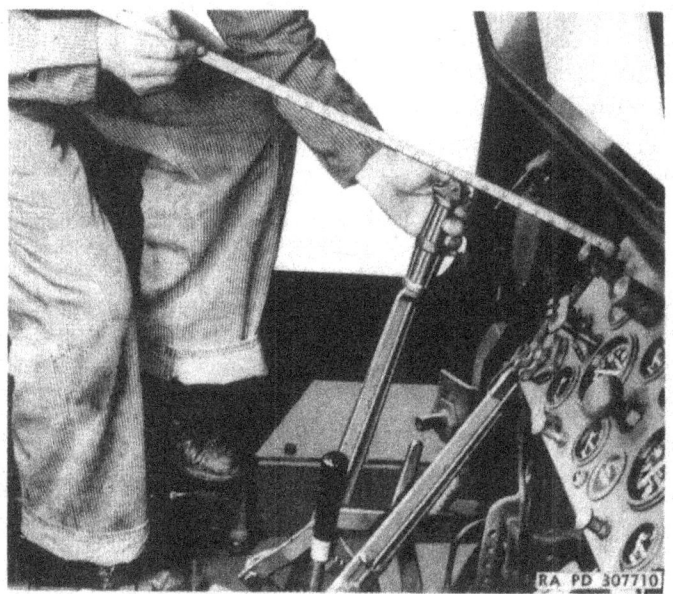

Figure 119—Measuring with Lever Pulled Back

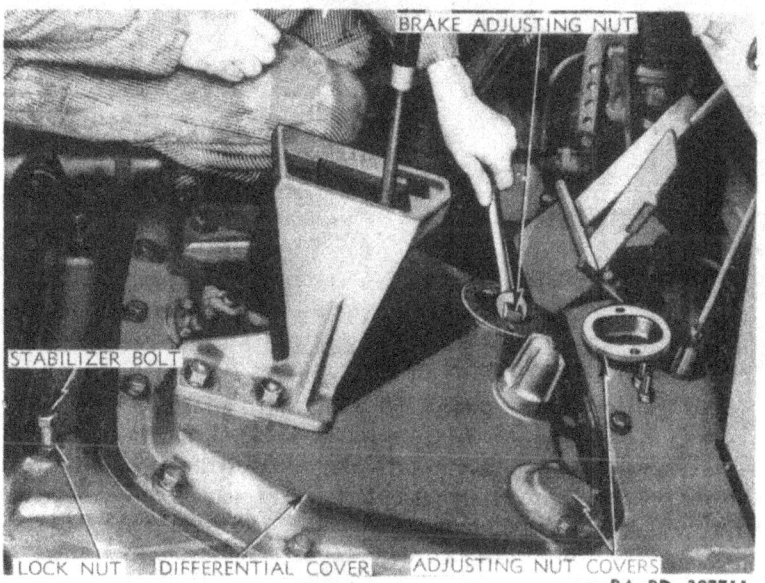

Figure 120—Adjusting Steering Brakes

18-TON HIGH SPEED TRACTOR M4

before engagement of brakes begins. Heating of transmission oil, and excessive wear on brake lining will result if brakes are too tight, and improper brake action will result if brakes are too loose or too tight.

b. **Adjustment Procedure.**

(1) CHECK ADJUSTMENT. Place scale or ruler against dash and over top of steering lever (fig. 118). Measure distance from dash to top of lever when lever is forward against stop. Hold scale in same position and pull back on lever until a slight pressure is felt and engagement of brake begins (fig. 119). The difference between the two lever positions should be 6 inches. If it is not, adjust as in (2).

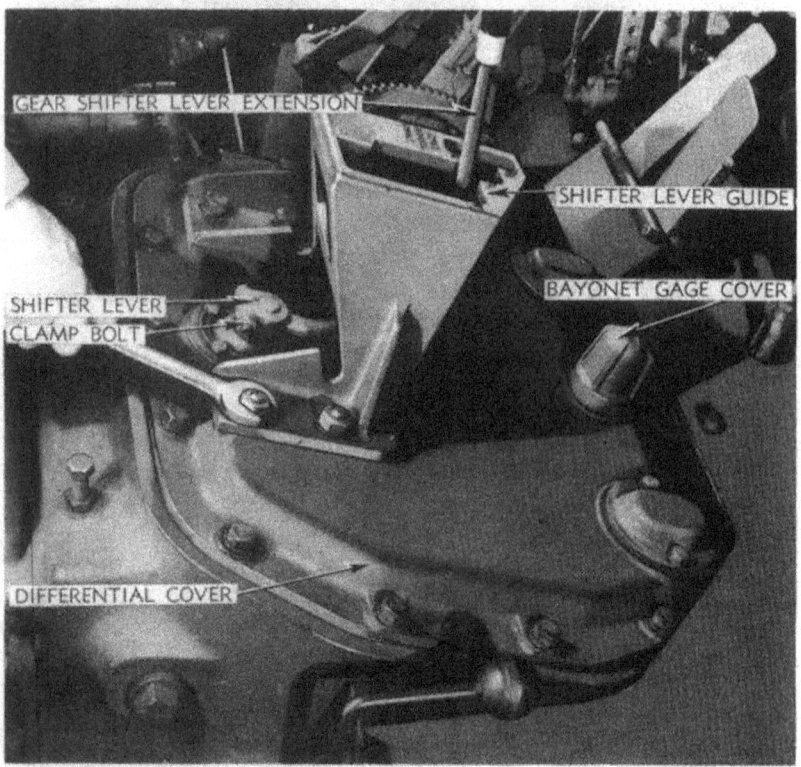

Figure 121—Removing Shifter Lever Guide

(2) ADJUST BRAKE BANDS. If free travel of lever is more than 6 inches, remove front seat cushions and seat plate and remove adjusting nut covers from differential cover (fig. 120) by removing two cap screws from each. Turn front adjusting nuts clockwise (if less than 6 inches, turn nuts counterclockwise) one-half turn at a time (fig. 120), testing free travel of lever after each time until

STEERING BRAKES

measurement is correct. With levers both adjusted to this measurement and forward against stop, loosen lock nuts and rear stabilizer bolts (fig. 120). Then turn bolts back in with fingers until pressure is felt as bolts contact support yokes. Turn bolts in one-half turn more and tighten lock nuts on bolts.

(3) INSTALL COVERS AND SEAT. Install adjusting nut covers over front adjusting nuts with two cap screws with lock washers in each cover. Install seat plate and cushions.

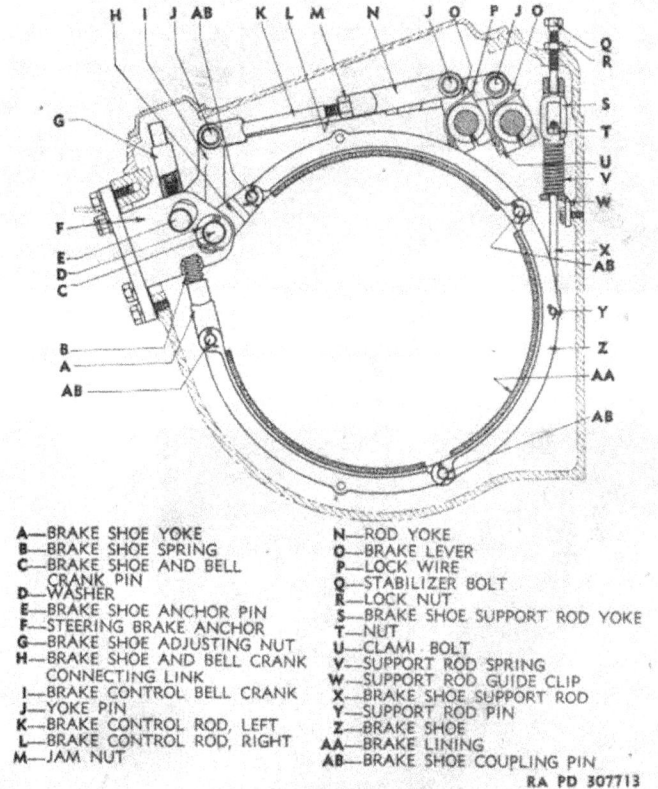

A—BRAKE SHOE YOKE
B—BRAKE SHOE SPRING
C—BRAKE SHOE AND BELL CRANK PIN
D—WASHER
E—BRAKE SHOE ANCHOR PIN
F—STEERING BRAKE ANCHOR
G—BRAKE SHOE ADJUSTING NUT
H—BRAKE SHOE AND BELL CRANK CONNECTING LINK
I—BRAKE CONTROL BELL CRANK
J—YOKE PIN
K—BRAKE CONTROL ROD, LEFT
L—BRAKE CONTROL ROD, RIGHT
M—JAM NUT
N—ROD YOKE
O—BRAKE LEVER
P—LOCK WIRE
Q—STABILIZER BOLT
R—LOCK NUT
S—BRAKE SHOE SUPPORT ROD YOKE
T—NUT
U—CLAMP BOLT
V—SUPPORT ROD SPRING
W—SUPPORT ROD GUIDE CLIP
X—BRAKE SHOE SUPPORT ROD
Y—SUPPORT ROD PIN
Z—BRAKE SHOE
AA—BRAKE LINING
AB—BRAKE SHOE COUPLING PIN

RA PD 307713

Figure 122—Steering Brake Assembly (Sectionalized View)

124. REPLACEMENT OF STEERING BRAKES.

a. **Removal of Brakes.**

(1) REMOVE DIFFERENTIAL COVER (fig. 121). Drain oil from differential case by removing cover from front hull guard and remove differential drain plug. Remove front seat cushions and seat plates. Remove two nuts and lock washers from right side of shifter lever

18-TON HIGH SPEED TRACTOR M4

Figure 123—Removing Brake Adjusting Nut

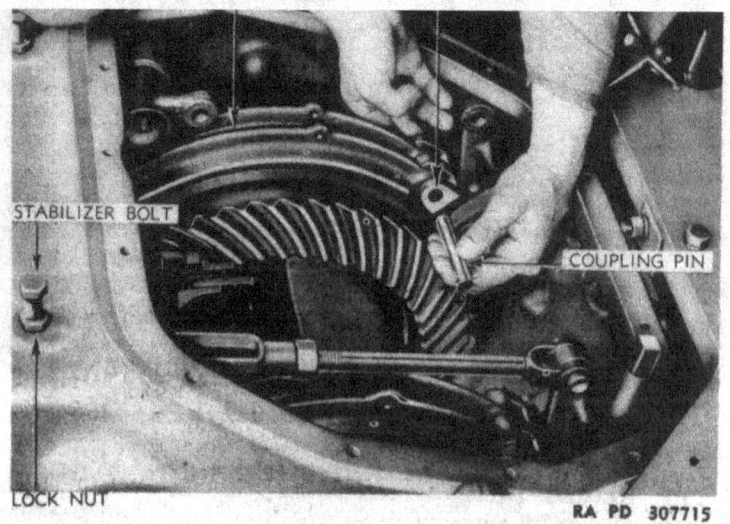

Figure 124—Removing Brake Shoe Coupling Pin

TM 9-785
124

STEERING BRAKES

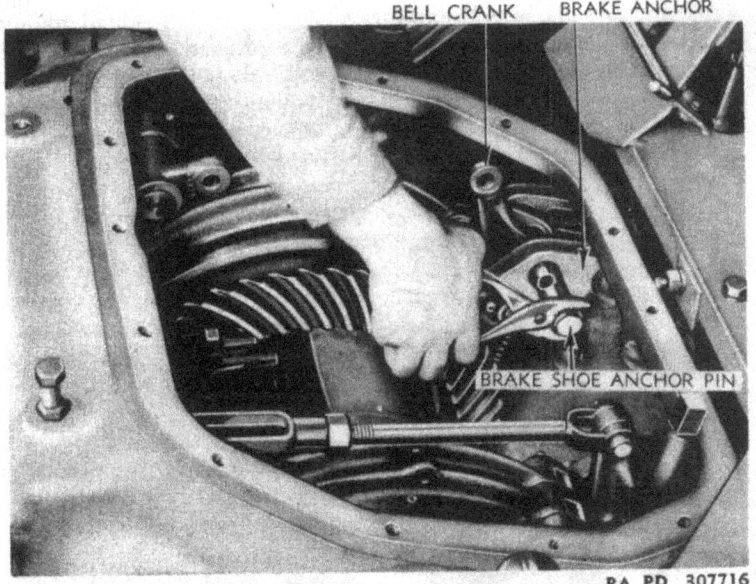

Figure 125—Removing Brake Shoe Anchor Pin

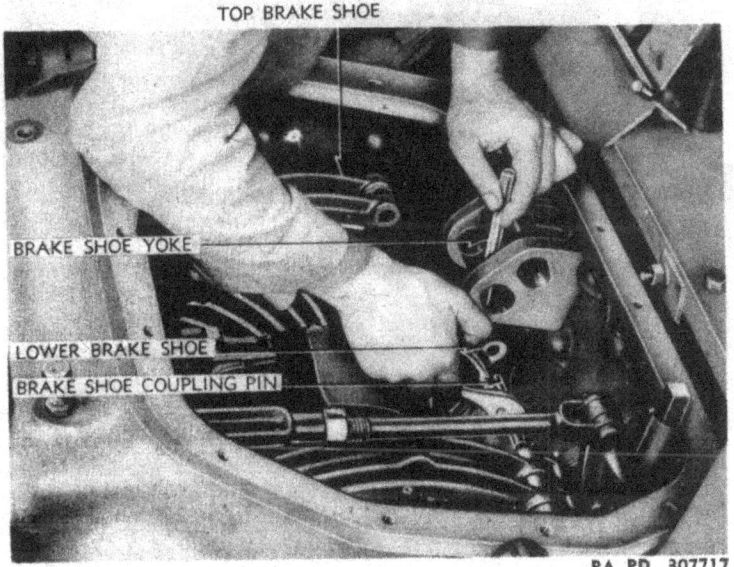

Figure 126—Removing Brake Adjusting Rod

TM 9-785
18-TON HIGH SPEED TRACTOR M4

guide and two cap screws and lock washers from left side of guide and remove guide. Loosen nut on clamp bolt on gear shifter lever extension and remove extension from lever. Unscrew bayonet gage cover from differential cover and pull out bayonet gage. Remove fourteen cap screws with lock washers from differential covers and remove cover.

(2) REMOVE BRAKE LINKAGE (figs. 122 and 123). Following procedure outlines removal of left brake assembly. Repeat each operation for right brake. Remove adjusting nut by unscrewing it from brake shoe yoke. Remove pin from yoke at front end of brake

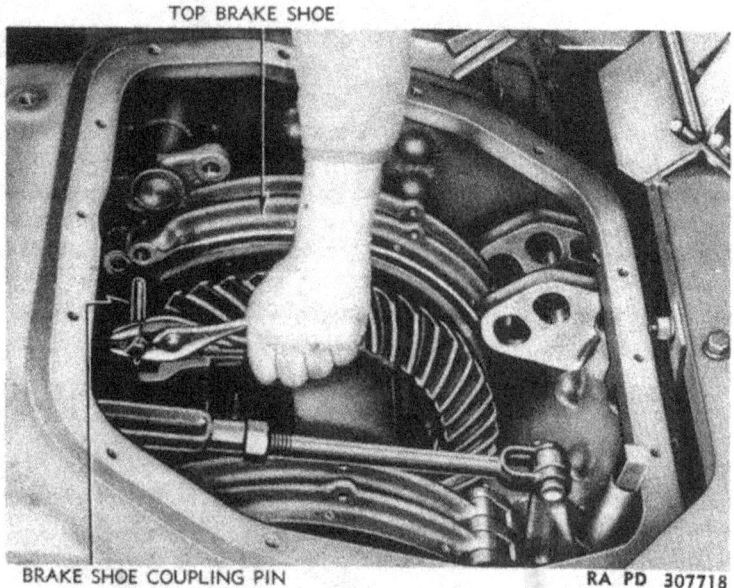

Figure 127—Disconnecting Upper and Center Brake Shoes

control rod connecting yoke to bell crank. Remove pin from yoke on rear end of rod and remove brake rod. Remove brake lever return spring from outside brake lever and bracket on power take-off housing (fig. 123). This spring must be removed in order to rotate brake shaft so head of clamp bolt may be turned out of groove in left brake shoe. It is not necessary for removal of right brake assembly. Remove pin from yoke to disconnect rear end of brake rod between bottom of left steering lever and lever on brake shaft.

(3) REMOVE BRAKE SHOES. Rotate brake shaft one-half turn so clamp screw is out of groove in left brake shoe. Remove pin from front end of top brake shoe and bell crank connecting link (fig. 124). Loosen lock nut and remove stabilizer bolt from brake shoe support

STEERING BRAKES

rod yoke. Pull cotter pin from nut inside brake shoe support rod yoke (fig. 122) and remove nut and yoke from support rod by turning nut off with yoke. Remove spring from rod. Rotate brake shoes forward around drum to drop brake shoe yoke out of brake shoe anchor pin, pull cotter pin, and remove anchor pin from anchor (fig. 125). Pull cotter pin and remove washer and large bell crank pin (lower pin) from bell crank connecting link and anchor and remove bell crank. Remove spring from brake shoe yoke. Pull cotter pin and remove coupling pin from brake shoe yoke and front end of lower brake shoe. Remove brake shoe yoke (fig. 126). Remove brake shoe coupling pin connecting top and center brake shoes (fig. 127) and remove top shoe. Rotate the two remaining shoes down around drum until support rod on center shoe slips out of rod guide clip, then remove the two shoes together (fig. 128). It will probably be necessary to bend support rod to slip it down out of guide clip on early tractors. Later tractors will have slotted clips. Straighten rod after brake shoes have been removed.

b. **Installation of Steering Brakes.**

(1) If brake shoe support rod guide clips (fig. 122) are not already slotted, remove clips by removing the two cap screws holding each clip to back of differential case. Cut slots through to holes for support rods wide enough for passage of rods. This will make it possible to install brakes without bending support rods (see a (3)). Install each clip with two cap screws and lock washers and thread lock wire through cap screw heads. On later tractors these clips will already be slotted.

(2) INSTALL BRAKE SHOE ASSEMBLIES. Procedure outlines installation of left brake. Repeat each operation for right brake. Couple center and lower shoes together with coupling pin and install brake shoe support rod on center brake shoe with support rod pin and cotter pin (fig. 128). Start end of lower shoe around back of drum (fig. 128) and slide shoes around drum until support rod slips into slot of rod guide clip. Lay third (top) shoe on brake drum and connect to center shoe with coupling pin (fig. 127). Install cotter pin in each end of pin. Install brake shoe yoke on front end of lower shoe with coupling pin through yoke and end of shoe. Secure coupling pin with cotter pins. Slip spring onto brake shoe yoke. Install brake bell crank in position shown in figure 123 and insert large brake shoe and bell crank pin through bell crank and anchor. Secure pin with flat washers and cotter pins at each end of pin. Insert brake shoe anchor pin in anchor (fig. 125), then rotate brake shoes around drum and slip brake shoe yoke up through hole in anchor pin. Slip spring onto brake shoe support rod above guide clip. Slip end of support rod yoke over support rod, then place castellated nut in yoke and turn nut onto rod by turning yoke (see fig. 122) until cotter pin can

18-TON HIGH SPEED TRACTOR M4

be inserted through nut and end of support rod. Install cotter pin. Screw stabilizer bolt through tapped hole in top of differential case

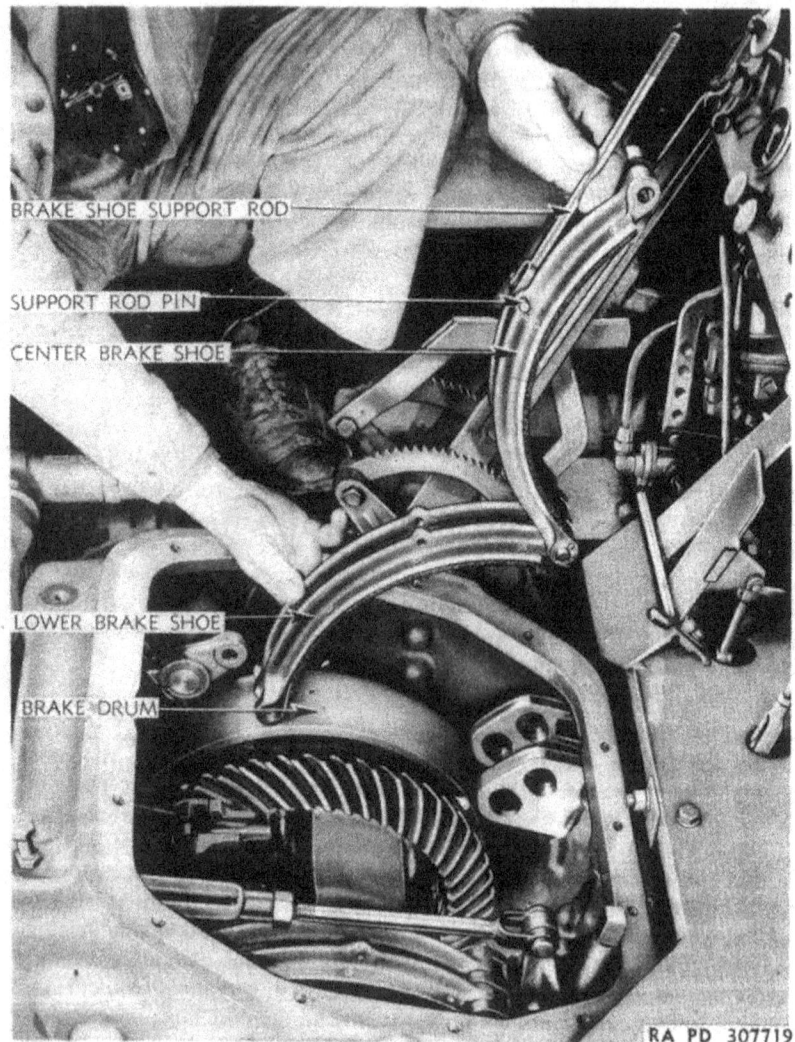

Figure 128—Installing Lower and Center Brake Shoes

until end of bolt starts into top of support rod yoke. Connect front end of top shoe to bell crank with coupling pin (fig. 124) and install cotter pin in each end of pin.

TM 9-785
124

STEERING BRAKES

(3) CONNECT BRAKE OPERATING LINKAGE (figs. 122 and 123). Adjust length of brake control rods (fig. 123) so distance between centers of holes in yoke and clevis for right brake is 14⅛ inches and for left brake 11¹⁵⁄₁₆ inches. Install rod, connecting ends to lever on brake shaft and bell crank with yoke pins as shown in figure 123. Install cotter pins in yoke pins. Screw adjusting nut onto brake shoe yoke. Install brake return spring (fig. 123), hooking one end of

Figure 129—Installing Differential Cover

spring into brake shaft lever, the other end into bracket on power take-off housing. Connect brake rod from steering lever to brake shaft lever with yoke pin and secure with cotter pin.

(4) INSTALL DIFFERENTIAL COVER AND GEAR SHIFTER LEVER GUIDE. Make sure cover gasket is in good condition or install new cover gasket. Lay cover on differential, engaging end of gear shifter lever in notches in shifter shaft extensions as cover is laid in place

233

18-TON HIGH SPEED TRACTOR M4

(fig. 129). Secure cover with fourteen cap screws with lock washers. Install lever extension on gear shifter lever in position shown in figure 121 and tighten clamp bolt. Install lever guide as shown in same figure with two cap screws and two nuts with lock washers.

(5) ADJUST BRAKES, INSTALL SEATS, AND FILL DIFFERENTIAL WITH LUBRICANT. Adjust brakes as outlined in paragraph 123 **b** (2), and install adjusting nut covers. Then place seats back in tractor. Install drain plug in bottom of differential case and install cover on bottom of hull guard with four cap screws. Fill differential with proper lubricant to level of "FULL" mark on bayonet gage. Install bayonet gage cover.

Section XXV

AIR TRAILER BRAKE CONTROLS

	Paragraph
Description of system	125
Air compressor	126
Air pressure governor adjustment	127
Air brake application valves	128
Tests for air brake system	129

125. DESCRIPTION OF SYSTEM.

a. Purpose. The purpose of the compressed air brake equipment on the tractor is to provide a means of operating the brakes on any vehicle that is equipped with air brakes and is being towed by the tractor and to provide a source of compressed air supply for other uses such as tire inflation. The air brake equipment has nothing to do with the application of the tractor brakes.

b. Components of Air Brake Control System. The Bendix-Westinghouse air brake control system used on the tractor includes the following units:

(1) An air compressor with a displacement of 7¼ cu ft per minute at its rated speed of 1250 revolutions per minute.

(2) An 8 x 26-inch air storage reservoir.

(3) A compressor governor which controls the air pressure in the air reservoir.

(4) A safety valve which protects the air system against excessive pressure.

(5) An air pressure gage to indicate the air pressure in the reservoir.

(6) Two brake application valves to control the air pressure delivered to the brakes of the towed vehicle.

(7) A double check valve to permit the use of two brake valves. This valve prevents air pressure from escaping through the exhaust valve of one brake valve when the other brake valve is used to apply the brakes.

(8) An air supply valve to provide a convenient means of tapping the compressed air supply on the tractor. This also permits higher air pressures to be obtained for such uses as tire inflation than that normally present in the air brake system.

(9) Two hose couplings mounted at the rear of the tractor for convenience in connecting the hose lines to the towed vehicle.

(10) Two cut-out cocks, also mounted at the rear of the tractor, which permit the connecting lines to be shut off when not in use.

TM 9-785
125
18-TON HIGH SPEED TRACTOR M4

(11) Two dummy couplings, also mounted at the rear of the tractor, which permit sealing the hose couplings against the entrance of dirt when they are not being used.

(12) A drain cock mounted in the bottom of the reservoir which permits the condensation which normally collects there to be drained. *This drain cock must be opened and reservoir drained daily.*

(13) Identification tags, mounted near the hose couplings at the rear of the tractor to identify the service line and the emergency line. Similar tags are mounted on all vehicles to permit easy identification of connecting lines.

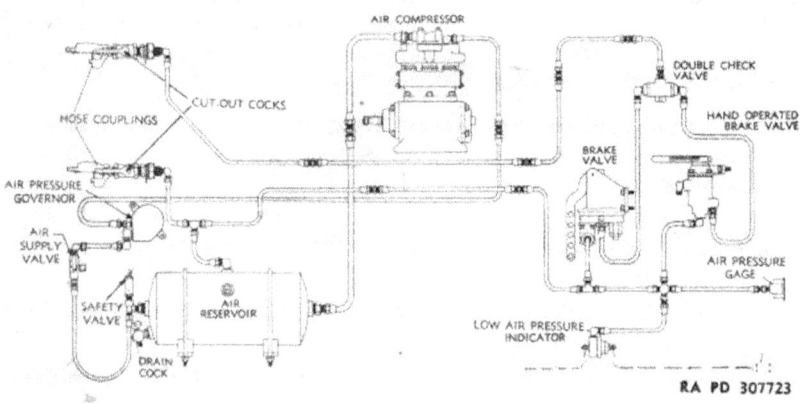

Figure 130—Schematic Diagram of Air Brake Control System

(14) Tubing and fittings which connect the various devices used in the air brake equipment.

c. Operation.

(1) During normal operation, the compressor which is controlled by the compressor governor, keeps the air pressure in the tractor reservoir between 85 pounds minimum and 105 pounds maximum. The tractor reservoir is connected to the air brake equipment on the towed vehicle through the emergency line. Therefore, the compressor keeps both air brake systems constantly charged. With both brake valves in released position, their inlet valves are closed and air pressure from the reservoir is prevented from passing through them. As

AIR TRAILER BRAKE CONTROLS

the driver depresses the foot pedal, the brake valve which is connected to the foot pedal is operated and its exhaust valve opens and its inlet valve closes. Air then flows through this brake valve to the double check valve and if necessary moves the shuttle in the double check valve to a position which blocks off the line leading to the other brake valve. This permits air pressure to flow through the double check valve and out the service line to the air brake system on the towed vehicle, applying the towed vehicle brakes. If the hand lever operated brake valve is used to apply the brakes, a similar action takes place.

(2) As soon as the air pressure in the service line corresponds to the position in which the driver has set the control mechanism of either brake valve, the inlet valve in the brake valve involved, closes. This prevents any further build-up of pressure in the service line and the brake valve is in its holding position. In other words, it is maintaining a pressure in the service line which corresponds to the amount the driver has moved the brake valve towards its fully applied position. When the driver partly releases the brake valve he has used to apply the brakes, the exhaust valve of that brake valve opens and permits air pressure in the service line to be reduced. If he moves the brake valve to full release position, all air pressure in the service line is exhausted.

(3) Before the brakes will function normally, the brake system on both the tractor and the towed vehicle must be charged to at least 70 pounds. The air pressure gage on the tractor records this pressure and the driver should always check to be sure this gage registers approximately 70 pounds before expecting maximum brake performance on the towed vehicle.

(4) When connecting any vehicle to the tractor, be sure the service line and emergency line connections at the rear of the tractor are properly connected to the service and emergency line connections on the towed vehicle. After this has been done, the cut-out cocks in the service and emergency lines at the rear of the tractor should be opened (handles parallel to lines). When opening or closing these cut-out cocks, the handles must always be turned with the hand and never struck with a hammer or some other heavy instrument, otherwise they may be damaged and leakage may develop.

126. AIR COMPRESSOR.

a. Description. The compressor is a two-cylinder, Bendix-Westinghouse, Model No. 2-UE-7¼-VW (fig. 131). It is mounted on a mat on the timing gear housing of the cylinder block. The compressor is water-cooled, is lubricated by oil from the engine lubrication system, and is driven by a belt from a pulley on the engine crankshaft. When the pressure reaches 105 pounds per square inch, the air pressure governor opens an unloader valve. This permits the air displaced by the

TM 9-785
18-TON HIGH SPEED TRACTOR M4

upward movement of the piston to pass from one cylinder to the other without compression until pressure drops to below 80 pounds, at which time more air is delivered to the reservoir.

b. **Maintenance of Air Compressor.**

(1) AIR RESERVOIR. The drain cock at bottom of air reservoir (fig. 130) must be opened daily and reservoir drained to remove con-

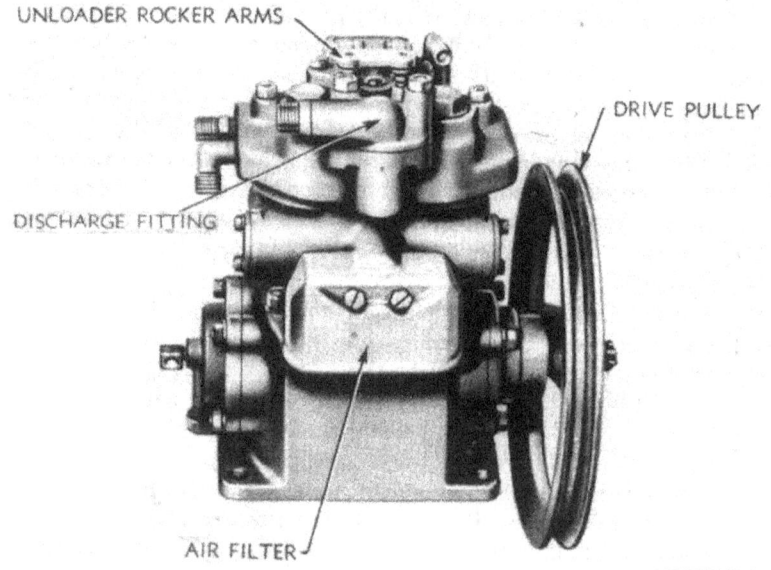

Figure 131—Air Compressor Assembly

densation which normally collects there. Be sure to close drain cock after draining reservoir.

(2) COMPRESSOR AIR FILTER.

(a) A dirty air filter will result in oil passing, loss of compression efficiency, and rapid wear. It is very important that air does not enter the compressor except through a clean filter. Filter and inlet chamber cover gaskets must, therefore, be kept in good condition and properly installed, otherwise, rapid wear in the compressor will result.

TM 9-785
125-126

AIR TRAILER BRAKE CONTROLS

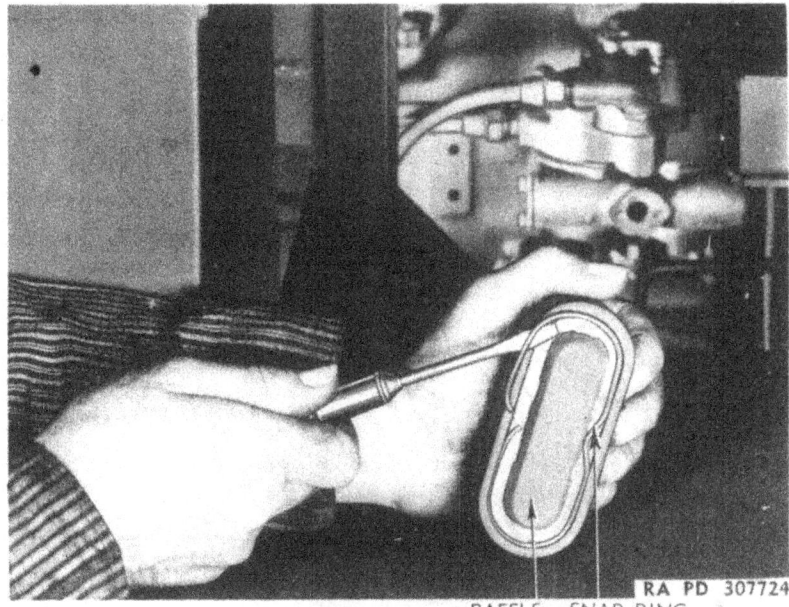

Figure 132—Removing Snap Ring from Compressor Air Filter

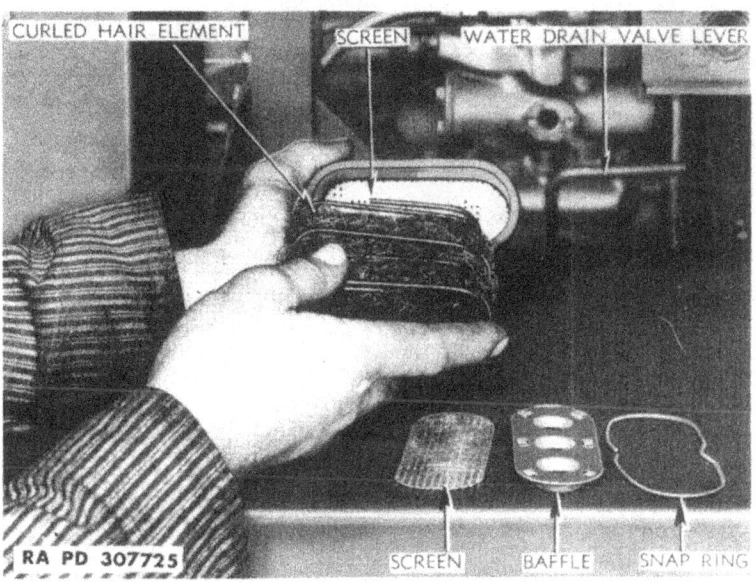

Figure 133—Removing Curled Hair from Filter

239

TM 9-785
18-TON HIGH SPEED TRACTOR M4

(b) To clean filter, remove the two cap screws holding filter to compressor and remove filter assembly. Pry snap ring from filter (fig. 132). Remove baffles, screens, and curled hair element from filter (fig. 133) and rinse hair and other parts in dry-cleaning solvent. Squeeze or shake dry-cleaning solvent from hair after rinsing and saturate with light engine oil. Place baffle, screen and hair back in filter, install second screen and baffle, snap ring, and install filter

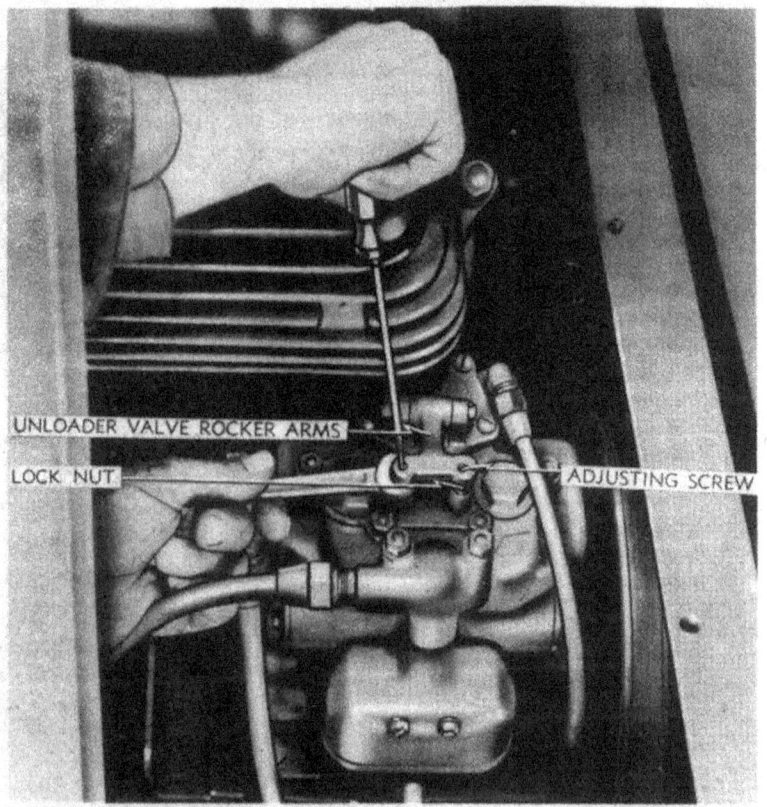

RA PD 307726

Figure 134—Adjusting Unloader Valve Clearance

assembly on compressor. (Later tractors will be equipped with an oil bath air cleaner.) Service and clean this oil bath cleaner at same intervals and in same manner as engine oil bath air cleaners.

(3) UNLOADING VALVES. Clearance should be 0.010 to 0.015-inch. Use feeler gage to check this clearance and adjust, if necessary, as follows: Loosen lock nuts on adjusting screws and turn screws to

TM 9-785
126

AIR TRAILER BRAKE CONTROLS

correct clearance (fig. 134). Turning screws counterclockwise increases the clearance; turning them clockwise decreases clearance. After correct clearance is obtained, tighten lock nuts. Then check clearance again to determine if clearance changed when lock nut was tightened. On later tractors, a cover is used to protect valves from dirt.

(4) COMPRESSOR DISCHARGE LINE. Remove compressor discharge line (fig. 132) periodically and clean it of carbon deposits or other forms of stoppage.

Figure 135—Removing Air Compressor Mounting Bolts

(5) DRIVE BELT. Adjust compressor drive belt periodically. Tighten belt by loosening lock nut and turning adjusting nut onto adjusting bolt of belt idler. Belt is adjusted correctly when straight side of belt may be deflected inward ¾ to 1 inch at point half-way between pulleys.

TM 9-785
126-127

18-TON HIGH SPEED TRACTOR M4

c. Air Compressor Replacement.

(1) REMOVE COMPRESSOR FROM TRACTOR. Open engine grille on left side of tractor. Drain cooling system. Disconnect the two water lines and two air lines from fittings on compressor head (fig. 134). Disconnect oil line at front end of compressor. Remove four cap screws holding air compressor to bracket (fig. 135), remove belt from pulley, and lift air compressor assembly from tractor.

(2) INSTALL REPLACEMENT UNIT. Loosen lock nut and adjusting nut on drive belt idler adjusting bolt. Set compressor in place on mounting pad on engine, place belt in drive pulley, and secure

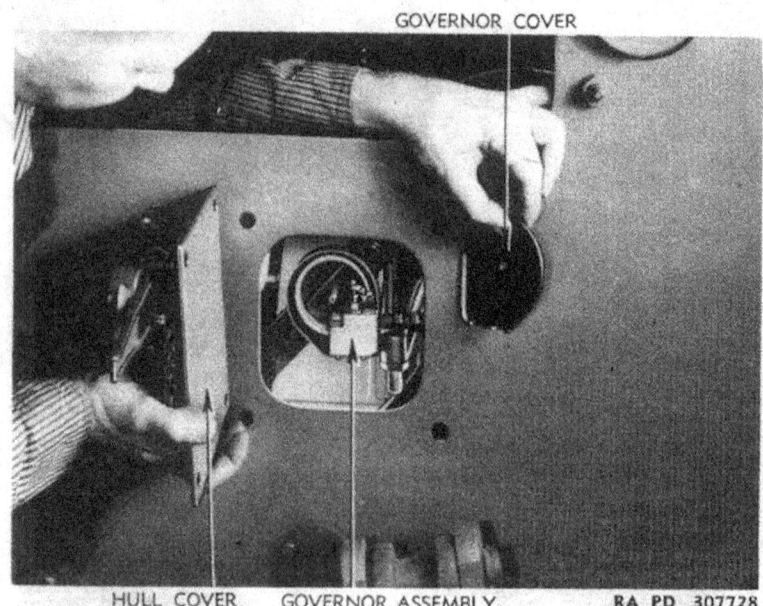

Figure 136—Air Pressure Governor Cover Removed

compressor with four cap screws with lock washers (fig. 135). Connect oil line to fitting on front end of compressor. Connect the two water lines and the two air lines (fig. 134) to fittings on compressor head. Adjust compressor drive belt for ¾ to 1-inch deflection. Tighten lock nut. Fill cooling system, then start engine and inspect for leaks at fittings.

127. AIR PRESSURE GOVERNOR ADJUSTMENT.

a. Air Compression. The governor should stop compression of air in air reservoir by compressor when pressure in reservoir reaches approximately 105 pounds and start compression when pressure drops

242

AIR TRAILER BRAKE CONTROLS

to approximately 85 pounds. Governor should be adjusted if it does not maintain pressures within this range. It is advisable to check the air pressure gage for accuracy by connecting an accurate gage in the air line before changing governor setting.

b. Adjustment Procedure. Operate engine and build up reservoir pressure. Observe on gage at what pressure the governor cuts out. Then slowly decrease pressure by opening reservoir drain cock or cut-out cock (fig. 9) and observe at what pressure governor cuts in. To change governor setting, remove cover in rear of hull by removing

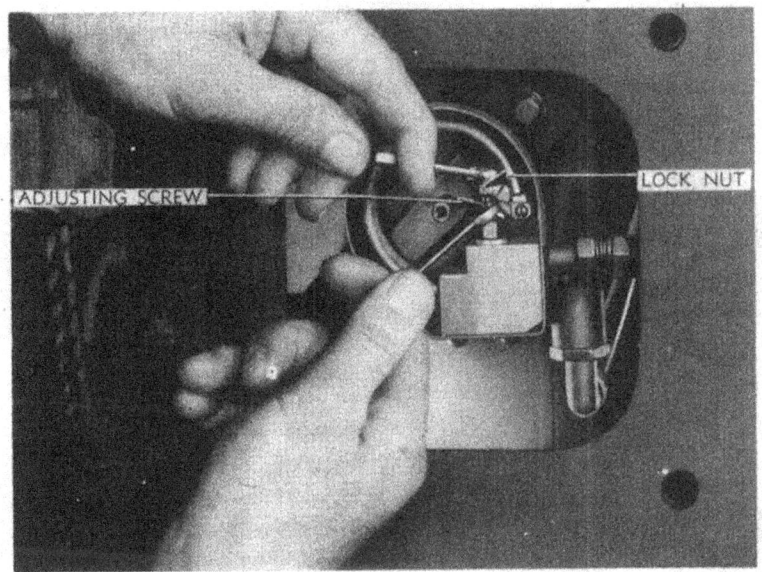

Figure 137—Adjusting Air Pressure Governor

four cap screws and lock washers, then remove screw from center of governor cover and remove cover (fig. 136). Loosen lock nut at top of adjusting screw (fig. 137). Turning adjusting screw clockwise will raise the cut-in and cut-out pressure settings and turning adjusting screw counterclockwise will lower them. After setting for proper operating pressure, tighten lock nut and replace covers.

128. AIR BRAKE APPLICATION VALVES.

a. The hand-operated brake application valve is mounted on dash at left of steering levers, the foot-operated valve controlled by the trailer brake pedal is mounted on bracket ahead of levers.

TM 9-785
128

18-TON HIGH SPEED TRACTOR M4

b. **Replacement of Hand-controlled Air Brake Valve.** Disconnect both inlet and outlet air lines from fittings on valve (fig. 138). Remove two cap screws holding valve on dash and remove valve assembly. Install replacement unit in its place, using reverse procedure. Start engine and check for leaks after building up pressure in reservoir.

c. **Replacement of Pedal-controlled Air Brake Valve.**

(1) REMOVE VALVE (refer to fig. 147). Remove electric brake controller as explained in paragraph 131 b. Pull cotter pin and remove yoke pin connecting control link to operating lever on valve. Discon-

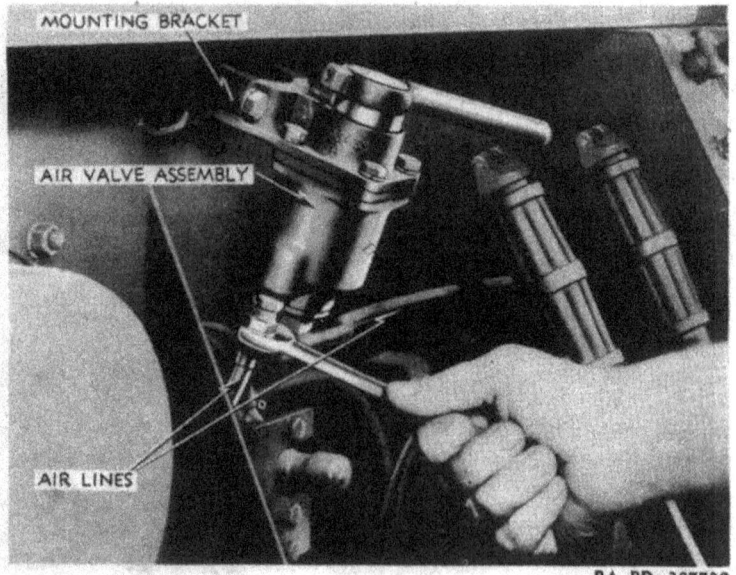

Figure 138—Disconnecting Air Lines from Air Brake Valve

nect all air lines from fittings on valve. Remove four bolts holding valve to bracket on dash and remove valve assembly.

(2) INSTALL VALVE ASSEMBLY (refer to fig. 147). Secure valve assembly to bracket on dash with four bolts with lock washers in position shown in figure 147. Connect air lines to fittings on valve. Connect control link to valve operating lever with yoke pin through rod yoke and lowest hole in lever. Install cotter pin in yoke pin. Install electric brake controller as outlined in paragraph •113 b. Operate engine to build up pressure and inspect connections for leaks.

(3) ADJUST LINKAGE. When trailer brake pedal is fully depressed, the air valve operating lever and operating lever on electric brake controller should move their full travel and contact their stops.

AIR TRAILER BRAKE CONTROLS

Adjust for full travel of air valve operating lever by removing end of control link from air valve lever, loosen lock nut, and shorten or lengthen link by turning it in or out of clevis. When correct adjustment is made, connect link to valve lever and install cotter pin in link pin. Tighten lock nut on links. Adjust for full travel of lever on electric brake controller in same manner.

129. TESTS FOR AIR BRAKE SYSTEM.

a. Frequency. Tests should be made at each inspection of the vehicle to make sure excessive leakage has not developed which

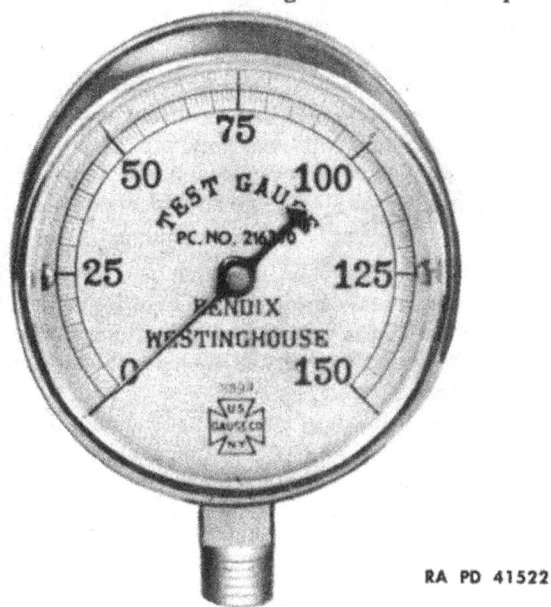

RA PD 41522

Figure 139—Air Test Gage for Checking Air Pressure in the Air Brake System

might impair the operation of the equipment and to make sure all devices are functioning normally.

b. Air Pressure Tests.

(1) With the motor running, observe at what pressure registered by the air gage the governor cuts out and compression is stopped. This pressure should be approximately 100 to 105 pounds.

(2) Observe at what pressure the governor cuts in and compression is resumed while slowly reducing the air pressure in the reservoir by applying and releasing one of the brake valves. This pressure should be approximately 80 to 85 pounds.

18-TON HIGH SPEED TRACTOR M4

c. **Leakage Tests.**

(1) With the brake system fully charged, the motor stopped and both brake valves in released position, observe the drop in reservoir air pressure registered by the air gage. The drop should not exceed 2 pounds per minute.

(2) With the motor stopped and both brake valves in applied position, observe the drop in reservoir air pressure registered by the air gage. The drop should not exceed 3 pounds per minute.

(3) Check for leaks in lines or connections with soapy water and a clean paint brush.

d. **Valve Delivery Pressure Test.** Connect an accurate air test gage to the service line outlet at the rear of the tractor and open the service line cut-out cock. When the foot-operated brake valve is depressed to its fully applied position, the air test gage should register approximately full reservoir pressure as registered on the dash gage. When the hand-operated brake valve is moved to fully applied position, the air test gage should register at least 60 pounds pressure.

e. **Operation Tests.**

(1) With the tractor connected to another vehicle with air brakes, test the operation of the foot brake valve by moving it to its applied position. Check the brakes on the towed vehicle to be sure that they apply and release properly.

(2) Test the operation of the hand brake valve by moving it to its applied position and check the brakes on the towed vehicle to be sure that they apply and release properly.

TM 9-785
130

Section XXVI

ELECTRIC TRAILER BRAKE CONTROLS

	Paragraph
Description of system	130
Replacement of control units	131

130. DESCRIPTION OF SYSTEM.

a. General. The Warner electric trailer brake control system consists of a resistor, load control, brake controller, and a coupling socket for attaching cables from the trailed unit to the tractor. These controls are for operation of the brakes on a trailed unit only and have nothing to do with the operation of brakes on the tractor.

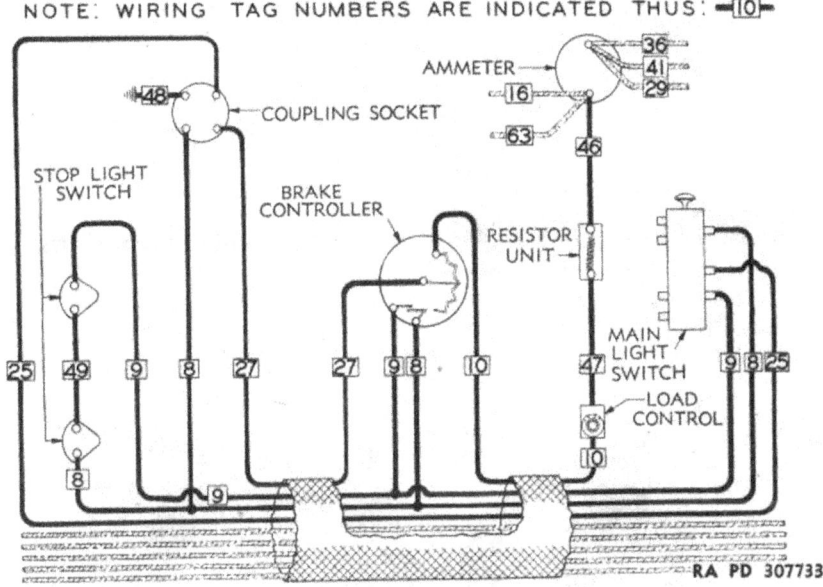

Figure 140—Wiring Diagram—Electric Trailer Brake Control System

(1) RESISTOR (fig. 141). The resistor reduces the 12-volt electrical system to the required 6 volts necessary to operate electric brakes. This unit is mounted in the instrument panel (fig. 96). A wire leads from one of the ammeter terminals to one terminal of this unit. Another wire leads from the other terminal of the resistor to the right-hand terminal of the load control (fig. 96).

(2) LOAD CONTROL (fig. 141). The load control, mounted on the instrument panel, allows the driver to regulate the power of the brake

247

TM 9-785
130

18-TON HIGH SPEED TRACTOR M4

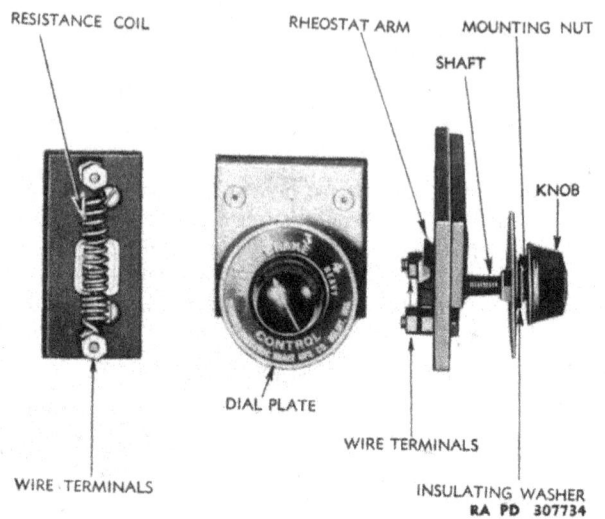

Figure 141—Electric Brake Resistor and Load Control Units

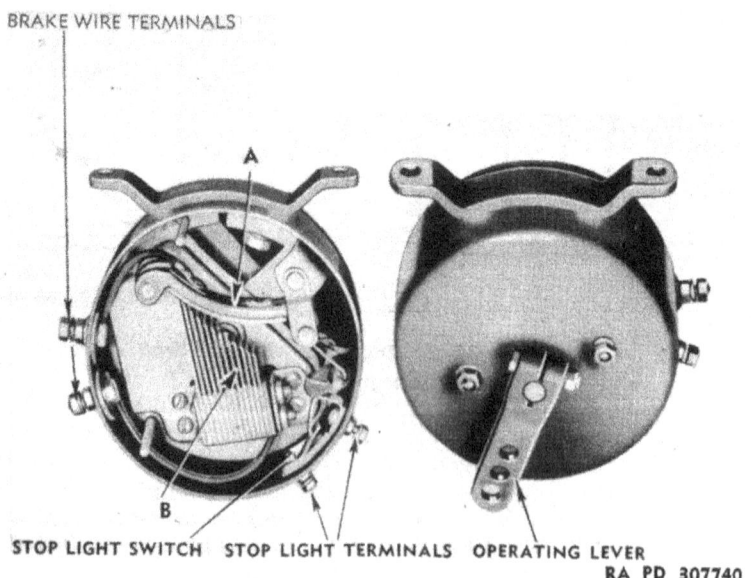

Figure 142—Electric Brake Controller

TM 9-785
130

ELECTRIC TRAILER BRAKE CONTROLS

to meet varying road or load conditions. A light trailed unit requires less severe application of the brakes than a heavy load. Severity of application is lessened by turning knob of load control to left (counterclockwise) and increased by turning knob to right (clockwise).

(3) BRAKE CONTROLLER (fig. 142). The brake controller is mounted on dash ahead of steering levers. Linkage from the trailer brake pedal is connected to a lever on the side of the controller (fig. 142). When the brake pedal is depressed, this moves the controller lever and brings the curved arm "A" (fig. 142) in contact with one or more blades "B." The least braking power is delivered when pedal is depressed only enough for curved arm to contact one blade. Maximum braking power is delivered when the pedal is depressed all

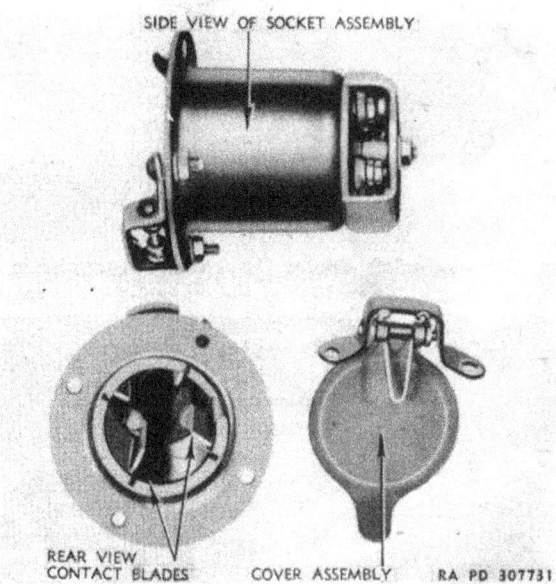

Figure 143—Electric Brake Coupling Socket

the way and the curved arm contacts all of the blades. The driver, therefore, can apply any degree of braking power desired. An integral stop light switch in controller (fig. 142) operates the trailer stop light when trailer brakes are applied.

(4) COUPLING SOCKET. The coupling socket is mounted at the rear of the tractor (fig. 148) and is designed for ready coupling of trailer brake cables. There are four terminals on the socket (fig. 149). The trailer taillight wire leading from the main light switch of tractor is connected to one; the stop light wire leading from brake controller is connected to another; the brake operating wire from controller

18-TON HIGH SPEED TRACTOR M4

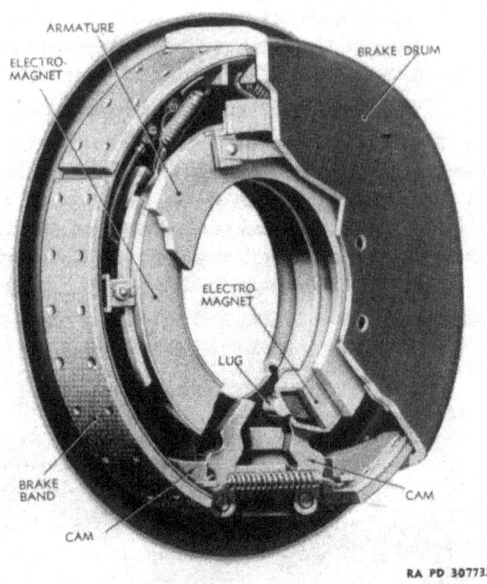

Figure 144—Electric Brake Operating Mechanism

Figure 145—Removing Load Control Mounting Nut

TM 9-785
130

ELECTRIC TRAILER BRAKE CONTROLS

Figure 146—Disconnecting Load Control Wires

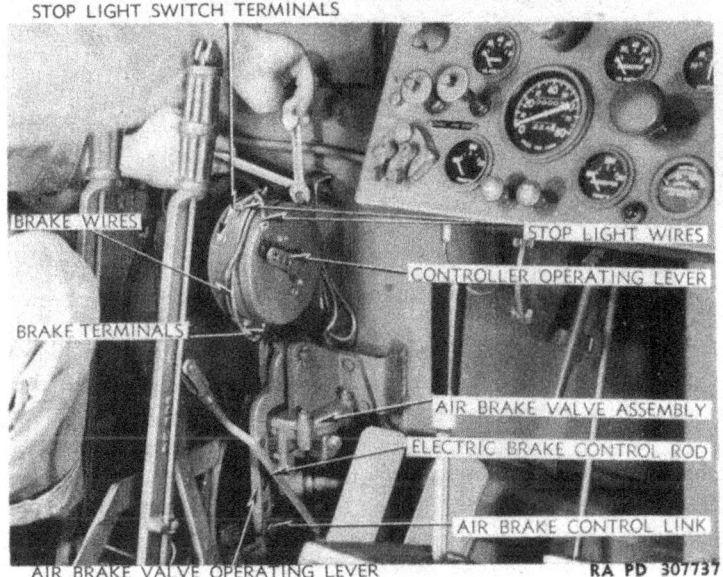

Figure 147—Removing Electric Brake Controller Mounting Bolts

TM 9-785
130

18-TON HIGH SPEED TRACTOR M4

is connected to the third terminal; and the ground wire from tractor frame is connected to the fourth terminal. When cable plug assembly from trailer is inserted into the coupling, the corresponding wires in trailer cable are automatically connected. The cover for the socket prevents entrance of dirt when electric trailer brake controls are not used.

b. **Operation of Electric Brake.** The electric trailer brake is a simple mechanical brake, operated by an electromagnet and armature disk in the trailer wheel (fig. 144). The armature revolves with the wheel. The magnet does not. As the lever on the brake controller on tractor is moved by depressing trailer brake pedal, thus making

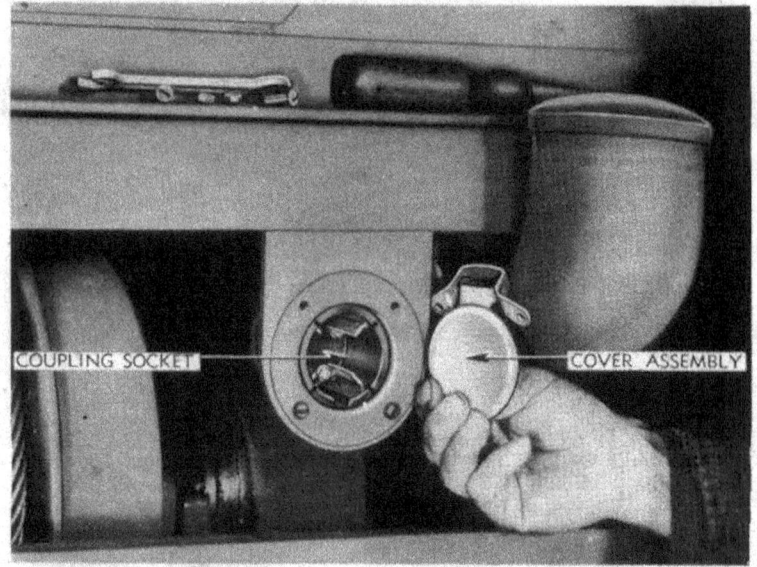

Figure 148—Coupling Socket Cover Removed

contact between the arm and blades in the controller, current flows through the electromagnet. The current energizes the magnet and causes it to cling to the revolving armature disk. The more current the driver allows to reach the magnet, the tighter it clings to the armature. This creates a drag on the magnet and causes it to shift in the direction of wheel rotation. When the magnet shifts, the attached lug presses one of the two cams (depending on direction of rotation) against the end of the brake band and forces the band against the brake drum. Grabbing does not occur because there is always a slight slipping action between the magnet and armature.

252

TM 9-785
131

ELECTRIC TRAILER BRAKE CONTROLS

131. REPLACEMENT OF CONTROL UNITS.

a. **Replace Resistor Unit** (refer to fig. 96). Disconnect instrument panel from dash as explained in paragraph 100 b (1). Remove nut from ammeter terminal and lift off end of wire leading to resistor. Remove nut from terminal on load control unit and lift end of wire leading to resistor. Remove two screws and lift resistor from panel. Install new unit by reversing above operations.

b. **Replace Load Control Unit.** Loosen set screw in side of knob and pull knob off shaft. Remove mounting nut (fig. 145), insulating washer, and dial plate from shaft, then remove remaining assembly from instrument panel. Remove nuts from terminals of unit (fig. 96) to disconnect wires. Connect wires to corresponding terminals of new unit, insert shaft through panel from rear, and install dial plate and insulating washer on shaft. Install mounting nut on shaft, turn

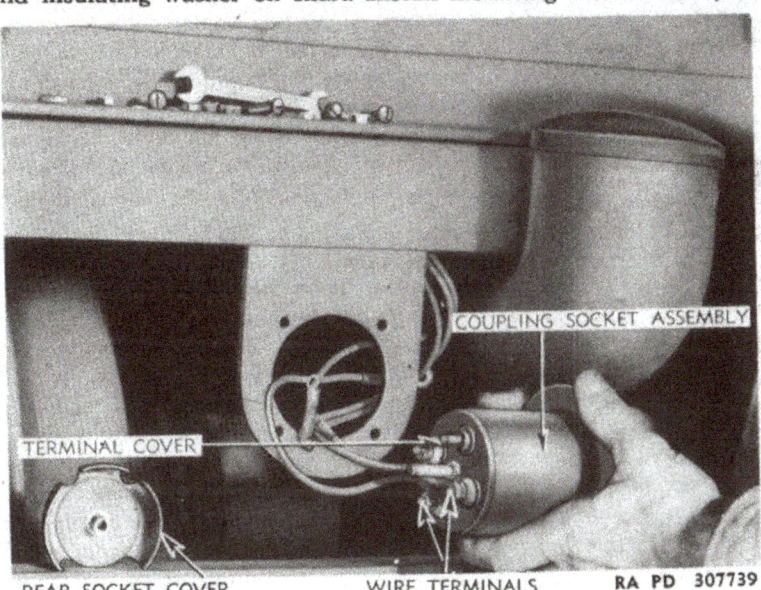

Figure 149—Coupling Socket Removed

dial to position shown in figure 145 and tighten nut. Slip knob on shaft and tighten set screw in knob against flat side of shaft.

c. **Replace Brake Controller Assembly.**

(1) REMOVE CONTROLLER (fig. 147). Disconnect trailer stop light wires from the two smaller terminals of controller. Disconnect brake wires from the two larger terminals. Pull cotter pin and remove pin from operating lever on controller and link. Remove two bolts holding controller to bracket on dash and remove unit.

253

18-TON HIGH SPEED TRACTOR M4

(2) INSTALL REPLACEMENT UNIT. Install replacement unit in reverse order to removal. Adjust lever travel on controller as outlined in paragraph 128 c (3).

d. **Replace Coupling Socket.** Remove two upper bolts and remove cover assembly from socket (fig. 148). Remove the two lower bolts and pull coupling socket assembly out of bracket (fig. 149). Remove nut and lock washer from center of cover and remove cover. Remove two nuts from each terminal, remove terminal covers, and lift off wires. Install wires on corresponding terminal of new unit as they are removed from old one. Install cover over terminals with nut and lock washer, and install socket in bracket. Install socket cover with four bolts with lock washers.

e. **Wiring System.** Refer to schematic wiring diagram (fig. 140) for wiring details of electric brake control system.

Section XXVII

TRACKS AND SUSPENSIONS

	Paragraph
General	132
Track assemblies	133
Track drive sprockets	134
Trailing idler assemblies	135
Bogie assemblies	136
Track support roller assemblies	137

132. GENERAL.

a. The weight of the tractor is carried by eight bogie wheels and the two trailing idlers. The tractor is supported on these wheels by brackets bolted to the sides and bottom of the tractor hull and equipped with volute springs to absorb road shocks when traveling over uneven terrain or rocks. The trailing idlers are adjustable to maintain correct track tension and volute springs in bracket allow idler wheel to oscillate on its pivot shaft. The bogie wheels and trailing idler wheels roll on the flat inner surfaces of the track shoes. The steel tracks which are driven by the drive sprockets bolted to the final drive hub are of standard ordnance design.

133. TRACK ASSEMBLIES.

a. **Description.** Each track is made up of 65 steel shoes with two rubber bonded pins pressed through them. These pins project from each side of shoes, and the shoes are connected to form a continuous belt by connectors bolted to the track shoe pins. These connectors also act as track guides. The teeth of the drive sprocket engage between the connectors to drive the track. When track connectors are worn on one side, they may be reversed so sprocket teeth will wear on other side of connectors, and thus more wear from connectors may be obtained. The tracks are not reversible and require no lubrication as there are no moving parts; flexing of track is allowed by rubber bushings on track shoe pins.

b. **Maintenance.** Keep wedges drawn tightly between track shoe pins. Check adjustment of tracks periodically as described in e and adjust as required. If track shoes appear to be wearing more on one end of the shoes than on the other, switch tracks to opposite sides of tractor so the side showing least wear will be on other side and thus compensate the wear. See that trailing idlers are kept in alinement to prevent excessive wear on track connectors and sides of wheels.

c. **Removal of Track Assembly.**

(1) LOOSEN TRACK. Remove nut lock from each side of idler over idler shaft nut after removing two cap screws from each (fig. 150).

18-TON HIGH SPEED TRACTOR M4

Loosen both idler shaft nuts and turn them back far enough for serrations on adjusting bracket and idler yoke to pass each other (fig. 151). Turn nuts back on track adjusting bracket bolts at sides of idler to allow idler to move forward (fig. 156) until nuts are even with ends of bolts.

(2) UNCOUPLE TRACK. Uncouple track at point between front bogie wheel and drive sprocket. Remove nuts from wedge bolts in two opposite track connectors (fig. 152) and drive wedges from connectors. Install a track connecting fixture (41-F-2997-86) on connector

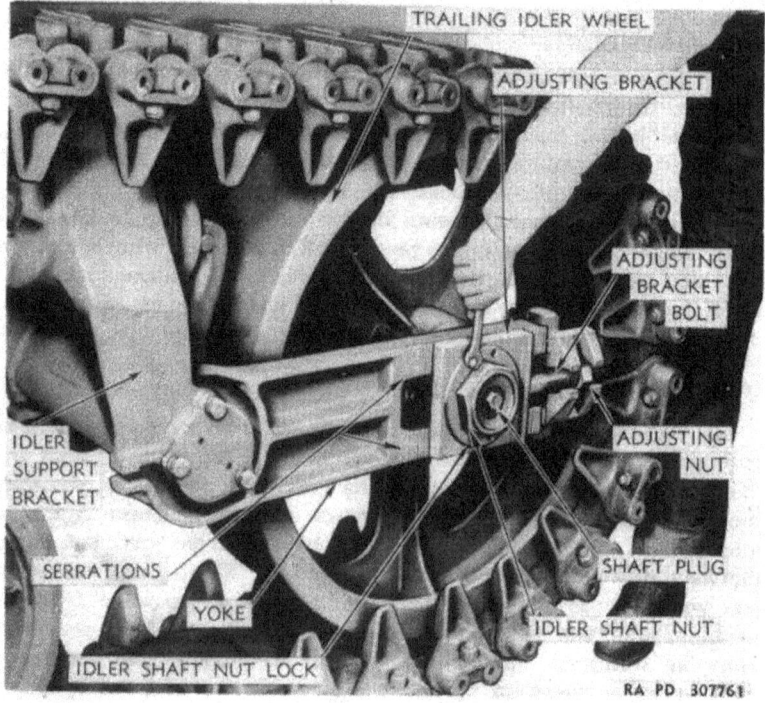

Figure 150—Removing Cap Screws from Nut Lock

on each side of track by inserting dowel on fixture into wedge bolt hole, then adjusting brace screw of fixture against top of connector to hold fixture level. Pins of fixture will rest against ends of track shoe pins. Turn screws of fixtures with handle (fig. 153) and force connectors off track shoe pins.

(3) REMOVE TRACK ASSEMBLY. Pull tractor backward with another machine until track end drops from trailing idler. If new track is to be installed, lay new track in straight line with old track and, with ends of two tracks against each other, pull tractor from old

TRACKS AND SUSPENSIONS

Figure 151—Loosening Trailing Idler Shaft Nut

Figure 152—Removing Nut from Wedge Bolt

TM 9-785
133
18-TON HIGH SPEED TRACTOR M4

track onto new track with another vehicle until front bogie wheel of tractor is resting on third shoe of new track. If tracks are to be switched from one side to the other, lay boards in place of new tracks and when tracks have been switched, pull tractor back onto tracks. Tracks may be completely disassembled by removing all track connectors.

d. **Installation of Track Assembly.**

(1) ROLL TRACK ONTO ROLLERS. After track has been laid down and tractor pulled onto track, hook rope to rear end of track, run rope over trailing idler, over support rollers between rollers and

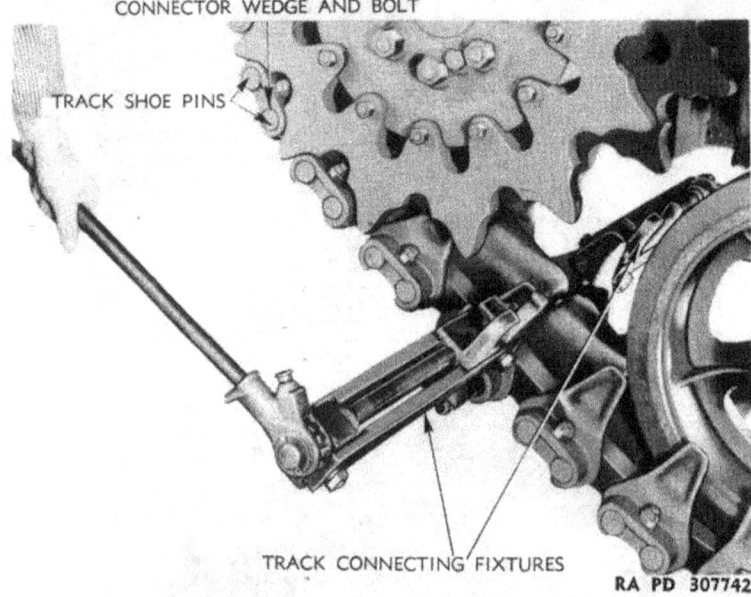

Figure 153—Removing Track Connectors with Fixtures 41-F-2997-86

over drive sprocket. Attach free end of rope to another machine and have this machine pull track up over trailing idler, support rollers and onto sprocket. Start track onto each roller with bar as it is pulled on. Roll sprocket forward with engine to roll track around it. Remove rope.

(2) CONNECT TRACK. Install a track connecting fixture (41-F-2997-86) on each side of track (fig. 154). Pull pins of shoes to be connected together with connector tools so that track connectors may be installed (fig. 154). Install a track connector on connecting pins and tap into place against ends of shoes. Install wedge in each of the two connectors, and start wedge nuts. Move tractor ahead until connectors are just started under front bogie wheel. Tighten nuts

TRACKS AND SUSPENSIONS

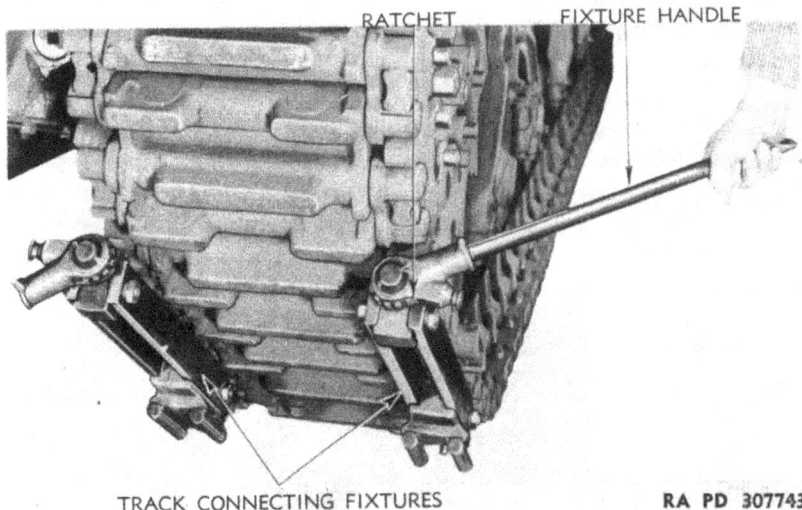

Figure 154—Coupling Track Assembly with Fixtures 41-F-2997-86

Figure 155—Checking Track Adjustment

TM 9-785
18-TON HIGH SPEED TRACTOR M4

securely. Remove track connecting fixtures and adjust track as outlined in e, this paragraph, "Track Adjustment."

e. **Track Adjustment.**

(1) LOOSEN IDLER SHAFT NUTS. Remove nut locks from trailing idler shaft nuts after removing two cap screws from each (fig. 150). Loosen idler shaft nuts (fig. 151) and back nuts off far enough to allow serrations in adjusting bracket to pass serrations in idler yoke.

(2) ADJUST TRACK TENSION. Track is correctly adjusted when it has 3/4-inch sag at a point half-way between track support rollers.

Figure 156—Tightening Track

Lay straightedge on top of track and measure from bottom of straightedge to top of track shoe (fig. 155). If track sags more than 3/4-inch at this point, turn nuts onto idler adjusting bracket bolts evenly (fig. 156) until correct measurement is obtained. Then tighten idler shaft nuts to draw adjusting brackets tightly against idler yoke with serrations of bracket entering serrations in yoke. Count serrations in yoke ahead of adjusting brackets on each side of yoke to be sure there are an equal number exposed on each side which will insure idler being in line with track.

TRACKS AND SUSPENSIONS

(3) INSTALL NUT LOCKS. Turn idler shaft nuts so holes in lock will line with holes in adjusting bracket, if necessary, and install nut locks with two cap screws with lock washers in each.

134. TRACK DRIVE SPROCKETS.

a. **Description.** Each track drive sprocket assembly consists of a hub bolted to the sprocket shaft and a sprocket bolted to each end of hub. The sprockets are reversible and may be changed from one side of hub to other to permit further use when one side of teeth becomes worn.

b. **Removal of Sprockets.**

(1) DISCONNECT TRACK. Follow procedure outlined under paragraph 133 c to uncouple tracks. Back tractor until track falls off

Figure 157—Removing Nuts from Sprocket Shaft Bolts

sprocket. Leave track intact on one side of tractor until sprocket is changed and track coupled again on first side.

(2) REMOVE SPROCKET ASSEMBLY. Remove eight high nuts from sprocket shaft bolts (fig. 157). Remove four puller hole cap screws (fig. 158) and install four puller cap screws in holes. Force sprocket assembly from shaft and bolts by turning puller cap screws in evenly (fig. 159). Remove thirteen bolts from each sprocket and hub and remove sprockets from sprocket hub (fig. 160).

18-TON HIGH SPEED TRACTOR M4

Figure 158—Removing Puller Hole Cap Screws

Figure 159—Pulling Sprocket Assembly from Shaft

TRACKS AND SUSPENSIONS

c. **Installation of Sprockets.**

(1) INSTALL SPROCKETS ON HUB. Bolt sprockets on sprocket hub with thirteen bolts with lock washers and nuts in each (fig. 160) with head of bolt toward inside. If worn sprockets are being installed, install each on side of hub that will cause track to wear on sides of teeth showing least wear. Either sprocket will go on either side of hub.

(2) INSTALL SPROCKET ASSEMBLY ON TRACTOR. Install sprocket over sprocket shaft and bolts (fig. 157) and install eight $7/8$-inch high nuts on bolts. Install four $3/4$-inch puller hole cap screws in threaded holes (fig. 158).

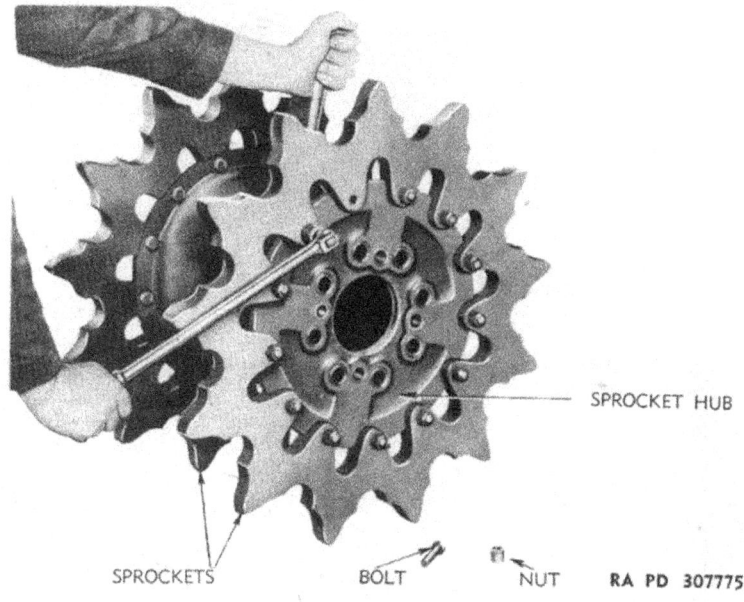

Figure 160—Removing Sprocket from Hub

(3) CONNECT AND ADJUST TRACK. Follow procedure outlined in paragraph 133 c to couple track and adjust track as outlined in paragraph 133 e.

135. TRAILING IDLER ASSEMBLIES.

a. **Description.** The trailing idlers are large steel wheels similar in construction to the bogie wheels. A volute spring in idler support bracket provides tension for the track and also allows idler and yoke to raise or lower when traveling over rough or rocky terrain. Serrations in the idler yoke and adjusting brackets provide for adjustment of track tension.

TM 9-785
135
18-TON HIGH SPEED TRACTOR M4

b. **Maintenance.** Keep idler wheels alined with tracks at all times to prevent excessive wear on track connectors and sides of idlers and bogie wheels. Track alinement depends on idlers.

c. **Removal of Idler Assembly.**

(1) UNCOUPLE TRACK. Refer to paragraph 133 c for procedure. Back tractor until end of track drops off idler to ground.

(2) REMOVE IDLER WHEEL ASSEMBLY. Remove idler shaft nuts (figs. 150 and 151). Pry adjusting brackets and bolts from each side of idler yoke. Remove two cap screws from each adjusting nut bracket (fig. 161) and remove brackets. Roll idler wheel assembly from yoke. Mark top side of square part of idler shaft bracket as wheel is rolled out.

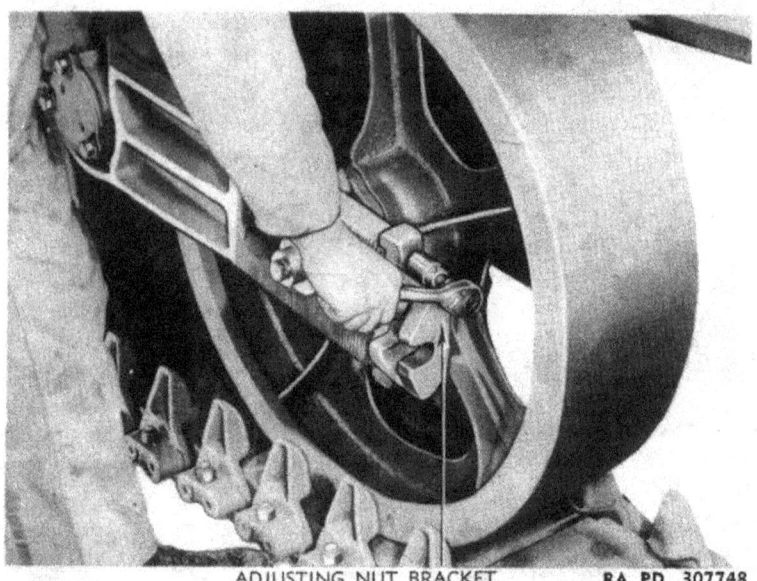

Figure 161—Removing Cap Screws from Adjusting Nut Bracket

d. **Remove Idler Yoke and Support Bracket Assembly.**

(1) When replacing either idler yoke or support bracket, it is better to remove the yoke while the entire assembly is bolted to tractor.

(2) REMOVE IDLER YOKE. Remove three cap screws and lock washers and remove cover from end of yoke pivot shaft (fig. 162). Screw adapter of special shaft puller (fig. 163) into tapped hole in end of shaft as tightly as possible, then screw bolt of pinion puller with sliding ram tightly into adapter. Using sliding ram, pull shaft from bracket (fig. 163) and remove yoke.

TRACKS AND SUSPENSIONS

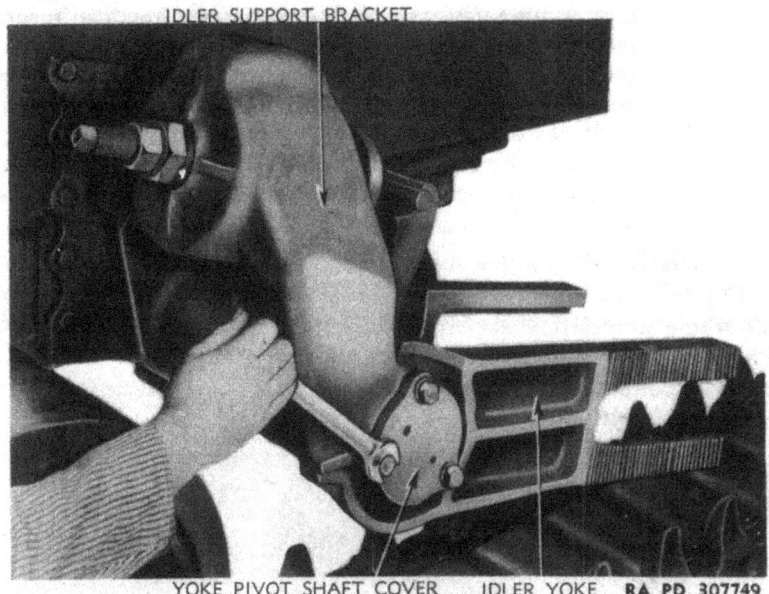

Figure 162—Removing Pivot Shaft Cover

Figure 163—Pulling Shaft from Idler Bracket and Yoke with Puller 41-P-2957-100

TM 9-785
18-TON HIGH SPEED TRACTOR M4

(3) REMOVE YOKE SUPPORT BRACKET. Remove four cap screws and lock washers and remove cover plate from rear end of side frame (fig. 164). Open grille on left side of tractor. Have another man hold nuts of bolts by reaching under ammunition box and through hole from which cover was removed (fig. 165) and remove the fifteen bolts attaching bracket to side frame. Remove bracket. Entire idler and bracket assembly can be removed as a unit by removing the above bolts.

e. Installation of Idler Assembly.

(1) INSTALL IDLER YOKE SUPPORT BRACKET. Install bracket on hull frame in position shown in figure 162 with fifteen bolts with lock washers. Install cover at rear of side frame (fig. 164) with four cap screws with lock washers.

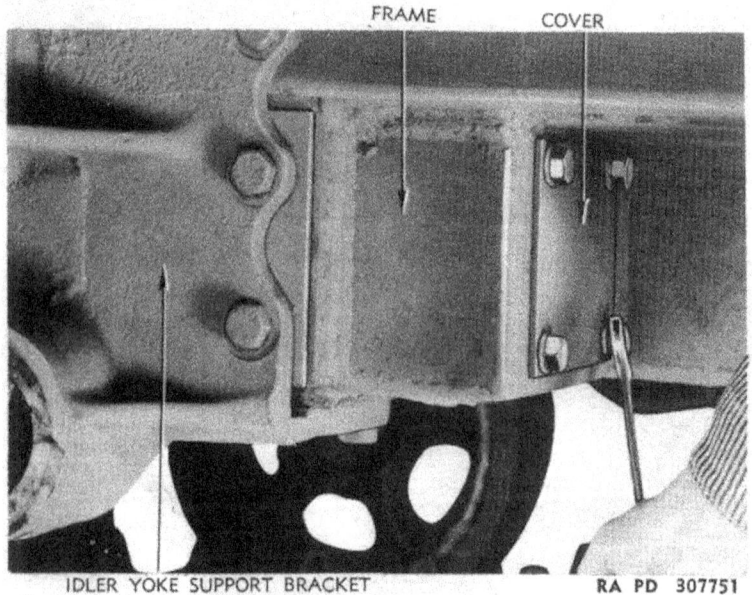

Figure 164—Removing Cover from Frame

(2) INSTALL IDLER YOKE. Have one man hold yoke with hole for shaft in line with holes in support bracket. Start end of shaft opposite end with dowel pins through bracket and yoke. Place cover (fig. 162) on dowel pins in end of shaft and aline holes in cover with holes in bracket for cover cap screws. Then tap shaft into place with heavy bar or hammer, driving against cover on shaft. Install cover over end of shaft with three cap screws with lock washers.

(3) INSTALL IDLER WHEEL ASSEMBLY. Roll idler wheel into yoke with end of shaft with plug to outside. Raise end of yoke and guide

TRACKS AND SUSPENSIONS

square parts of idler shaft brackets into slides in yoke. NOTE: *Only two sides of square part of bracket are machined, these two sides must be at top and bottom when installed (mark on bracket and mark on lubrication hole end of shaft toward top).* Install an adjusting nut bracket on end of each side of yoke with two cap screws with lock washers (fig. 161). Place an adjusting bracket and bolt on each end of idler shaft with nuts on bolts entering slot in brackets just installed. Start idler shaft nuts onto ends of shaft, leaving them loose enough for serrations to pass each other when adjusting track.

(4) INSTALL AND ADJUST TRACK. Follow procedure outlined in paragraph 133 d and e to install and adjust track assembly.

Figure 165—Removing Idler Yoke Support Bracket

136. **BOGIE ASSEMBLIES.**

a. Description. There are two bogie assemblies on each side of tractor, consisting of two rubber-tired steel wheels held in brackets which oscillate on pivot shafts in the bracket that is bolted to hull. A volute spring between the upper parts of the two bogie wheel brackets absorbs road shocks and allows wheels to raise or lower a limited distance when traveling over rough or rocky ground. The bogie wheels rotate around their axle shafts on tapered roller bearings and are equipped with positive type oil seals to prevent leakage of oil or entrance of dirt.

18-TON HIGH SPEED TRACTOR M4

b. Maintenance.

(1) LUBRICATION. Use the special flushing lubricator to change oil used in the bogie wheels (see lubrication instructions). This method consists of forcing the old oil out of the wheel, which also flushes dirt and foreign material out with it, and filling the wheel with new lubricant.

(2) Keep all bolts tight and tracks and trailing idlers alined to prevent wear on sides of wheels and tires. Keep dirt and mud cleaned from wheels and brackets as much as possible especially when it is likely to freeze.

c. Replacement of Bogie Wheel. Place jack at side of track and jack up under bogie arm until bogie wheel is lifted off track. Remove

Figure 166—Removing Bogie Wheel

four cap screws holding clamps to arm and remove caps. Roll bogie wheel clear of arm and lift out. Install replacement bogie wheel assembly, using reverse procedure, being sure that mark on lubricating hole end of shaft is up.

d. Removal of Bogie Components. The following procedure outlines removal of complete assembly and separation into component parts. It is not necessary to uncouple the track to remove the bogie assembly, however, in the accompanying illustrations, the track was uncoupled so that operations could be more clearly shown.

TRACKS AND SUSPENSIONS

Figure 167—Removing Volute Spring Seat Caps

Figure 168—Compressing Volute Spring

18-TON HIGH SPEED TRACTOR M4

(1) BLOCK UP UNDER HULL FRAME. Place jack under end of side frame on side from which bogie assembly is to be removed, and raise frame slightly. Block up under frame and let weight of tractor down onto blocks.

(2) REMOVE VOLUTE SPRING ASSEMBLY. Remove sixteen cap screws holding the four volute spring seat caps to bogie arms (fig. 167) and remove caps. Insert volute spring compressor bolt through spring seats and spring. Install nuts on compressor bolt and turn nuts onto bolt (fig. 168) until spring is compressed far enough that spring and caps may be lifted from bogie arms. Lift volute spring

Figure 169—Removing Volute Spring and Seats

and seats from arms (fig. 169). Remove nuts from volute spring compressor and separate spring and spring seats.

(3) REMOVE BOGIE ARM ASSEMBLIES. Remove three bolts holding shaft clamp to outer ends of pivot shafts (fig. 170). Remove clamp. Remove inner cap screw, and loosen two outer cap screws in shaft clamp of support bracket holding pivot shaft of bogie arm to be removed (fig. 171). Spread clamp slightly with chisel to free shaft. Block up under bogie wheel so it will clear track guides, pry bogie assembly from support bracket and remove bogie assembly from tractor (fig. 172). Repeat above operations if support bracket is to be removed.

TRACKS AND SUSPENSIONS

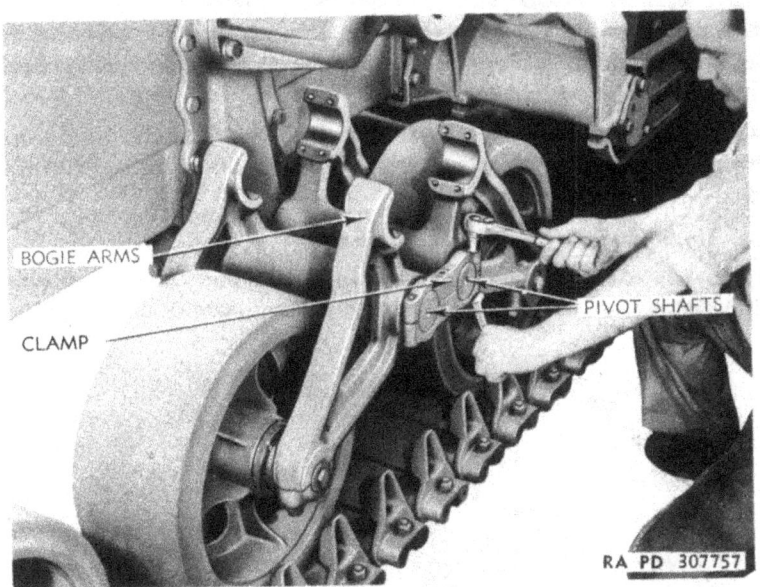

Figure 170—Removing Bolts from Pivot Shaft Clamp

Figure 171—Loosening Clamp Cap Screws

TM 9-785

18-TON HIGH SPEED TRACTOR M4

(4) REMOVE BOGIE SUPPORT BRACKET. Remove track support roller assembly by removing six cap screws. Remove seven cap screws and six bolts holding bogie bracket to hull and remove bracket and spacers between bracket and hull.

(5) REMOVE WHEELS FROM BOGIE ARMS. Remove two cap screws from each cap and remove caps holding wheels in bogie arms. Remove wheels from arms.

e. **Installation of Bogie Components.**

(1) INSTALL SUPPORT BRACKETS. Install support bracket on hull in position shown in figure 172 with seven cap screws and six bolts

Figure 172—Bogie Wheel and Arm Removed

with lock washers. Use spacers between bracket and bottom of side of hull. Install track support roller assembly on support bracket in position shown in figure 172 with six cap screws with lock washers.

(2) INSTALL BOGIE ARM ASSEMBLIES. Start ends of pivot shafts in bogie arms into support bracket and drive shafts into bracket. Do not tighten clamps. Place bogie wheels in bogie arms and install lower caps on arms around bogie wheel shafts with two cap screws with lock washers in each cap (fig. 166). Install clamp (fig. 170) on ends of pivot shafts with three bolts with lock washers. Place a spring seat on each end of volute spring and compress spring enough

TRACKS AND SUSPENSIONS

with volute spring compressor bolt to lay spring assembly in upper parts of bogie arms (fig. 169). Remove compressor bolt and install the four caps over ends of spring seats with four cap screws with lock washers in each (fig. 167).

(3) LOWER TRACTOR AND TIGHTEN SUPPORT BRACKET CLAMPS. Raise tractor with jack and remove blocks, then lower tractor with jack so weight of tractor will cause bogie pivot shafts to assume their natural position. Then install inner cap screw with lock washer in inner hole of clamp at bottom of support bracket. Tighten the three clamp cap screws for each pivot shaft (fig. 171).

Figure 173—Removing Track Support Roller

137. TRACK SUPPORT ROLLER ASSEMBLIES.

a. **Description.** The track support rollers (two at each side) are double-wheeled rollers which support the track as it returns from the trailing idler to the drive sprocket. The roller support bracket mounts on the same bracket on hull as the bogies. The rollers turn on tapered roller bearing and are lubricated in same manner as the bogie wheels and trailing idlers. Spring loaded oil seals protect roller against entrance of dirt or leakage of oil.

b. **Maintenance.** Keep mounting bolts of roller assemblies tight. Lubricate rollers after every 8 hours of operation as outlined in lubrication instructions.

18-TON HIGH SPEED TRACTOR M4

c. **Replacement of Support Roller Assembly.** Place jack on bogie wheel tire directly beneath support roller assembly to be removed and raise track off of support roller. Remove six cap screws and lock washers holding support roller assembly to bogie support bracket (fig. 173) and remove support roller assembly. Reverse operations to install roller assembly.

Section XXVIII

WINCH AND POWER TAKE-OFF

	Paragraph
Description	138
Maintenance	139
Winch adjustments	140
Replacement of winch assembly	141
Replacement of winch drive shaft assembly	142
Replacement of power take-off assembly	143

138. DESCRIPTION.

a. **Winch.** The winch is a Gar Wood, Model No. 4M 718, with 300 feet of ¾-inch cable. It is of the standard heavy duty military type, and mounted in rear end of tractor hull. Drive shafts with universals are connected to the winch worm shaft in winch gear case and to the power take-off which operates the winch. The winch drum shaft is driven by the worm and gear in gear case through a sliding jaw clutch operated by the control lever in driver's compartment (fig. 5). This worm and gear reduction provides for maximum pull on cable. An automatic brake on the winch worm shaft holds loads suspended when engine master clutch is disengaged.

b. **Power Take-off.** The power take-off is mounted on the transmission case and driven by a gear on the transmission shaft. It is of the reversible type and provides for turning winch drum in either direction to wind cable onto drum, or unwind cable from drum. A sliding pinion in the power take-off permits shifting the power take-off into winding, unwinding, or neutral position. It is operated by the control lever located between the steering levers (fig. 5).

139. MAINTENANCE.

a. Lubricate winch and power take-off as directed in Lubrication Guide and instructions. Keep winch sliding jaw clutch adjusted so jaws on clutch mesh fully with jaws on winch drum. Keep winch worm automatic safety brake adjusted (par. 140 a) so it will hold load suspended when engine master clutch is disengaged.

140. WINCH ADJUSTMENTS.

a. **Adjustment of Automatic Brake.**

(1) REMOVE COVERS. Remove eight cap screws and remove plate over opening in rear of hull (fig. 174). Remove three wing nuts and one wing bolt from upper brake housing covers and remove covers (fig. 175).

TM 9-785
141-142

18-TON HIGH SPEED TRACTOR M4

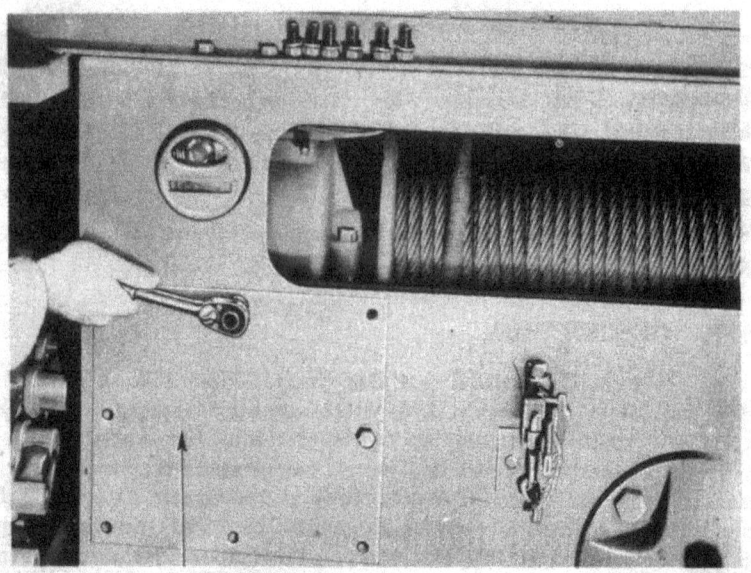

Figure 174—Removing Plate from Hull Opening

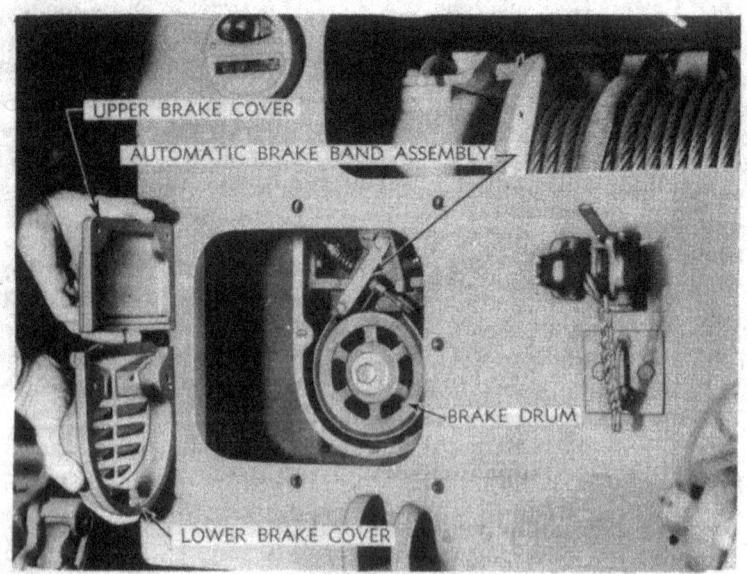

Figure 175—Brake Housing Covers Removed

TM 9-785
140

WINCH AND POWER TAKE-OFF

(2) ADJUST BRAKE.

(a) If brake is being installed after having been removed, wedge rocker over so that rocker holds contact at "C" (fig 177). Slack off top lock nut and adjusting nut about ¼ inch and adjust top adjusting nut to make length of brake spring in this position 1⅝ inch. (dimension "A"). Tighten lock nut against top adjusting nut until there is ⅛-inch clearance between spring spacer and lower adjusting nut. Test brake now to see if it will hold load. If it does not, increase brake spring tension as outlined in following steps.

(b) Loosen lock nuts above top brake adjusting nut and lock nut below lower brake adjusting nut. Tighten top adjusting nut ½ turn (fig. 176). Tighten lock nut against it to maintain location.

Figure 176—Adjusting Automatic Brake

(c) With no load on winch, wedge rocker over so that rocker holds contact at "C" (fig. 177). Use screwdriver to hold spring spacer up into rocker as far as it will go.

(d) Dimension "B," which is the space between spring spacer and lower adjusting nut, should then be about ⅛ inch (fig. 177). Tighten or loosen (whichever is necessary) lower adjusting nut until dimension "B" is about ⅛ inch. Then maintain location of lower brake adjusting nut by tightening lock nut against it.

Figure 177—Checking Adjustment of Automatic Brake

WINCH AND POWER TAKE-OFF

(e) Test winch. If brake does not stop load and hold it, tighten top adjusting nut again, by ½ turn, then retest winch. Keep repeating this procedure until brake holds load.

(f) If brake overheats during hoisting operation, slightly decrease dimension "B" until heating becomes normal. To decrease dimension "B" (fig. 177), back off lock nut of lower adjusting nut, then turn lower adjusting nut clockwise, by ½ turn, and tighten lock nut against it.

(g) If brake overheats in winch lowering operation, slightly decrease tension of brake spring until heating becomes normal. First back off lock nut of top brake adjusting nut and then back off top adjusting nut ½ turn. Lock position of top adjusting nut by tightening lock nut against it.

(h) Test brake by running winch with no load in lowering position. If overheating is not corrected, repeat procedure until overheating is stopped.

(i) To determine whether or not overheating is caused by hoisting or lowering operations, run winch first in hoisting operation, then in lowering operation. In hoisting operation, brake should not heat even if run continuously. In lowering operation, there should be no excessive heating, although brake will become too hot to touch. NOTE: *Smoking of brake does not necessarily indicate improper adjustment, as oil on brake disk or lining will smoke until burned off, and this will burn off at much lower temperatures than will harm lining. If lining chars, heating is excessive and adjustment should be made.*

(3) INSTALL COVERS. Install brake housing covers with three wing nuts and one screw. Install cover over opening in hull with eight cap screws and lock washers.

b. Adjustment of Winch Jaw Clutch. The sliding jaw clutch must be kept adjusted so that when the clutch lever is in engaged position, the jaws on clutch will mesh fully with jaws on winch drum; also when winch clutch lever is in disengaged position, the jaws on clutch must disengage fully from jaws on drum. Remove pin from clevis connected to lever on winch clutch operating shaft, loosen jam nut and turn clevis on control rod to shorten, or off to lengthen rod to correct mesh of clutch jaws with drum jaws.

141. REPLACEMENT OF WINCH ASSEMBLY.

a. Remove Ammunition or Cargo Box. Refer to paragraph 144 or 145 for removal procedure.

b. Disconnect Winch Clutch Control Rod and Drive Shaft. Pull cotter pin and remove yoke pin connecting winch clutch control rod to lever on winch clutch operating shaft. This will also allow re-

TM 9-785
141-142

18-TON HIGH SPEED TRACTOR M4

moval of spacer on pin. Disconnect spring. Remove hexagon recessed set screw from end of universal on winch worm shaft (fig. 178).

c. Remove Winch Assembly from Tractor. Remove six cap screws holding winch base to tractor hull. Tap rear drive shaft forward until universal is driven off worm shaft. Fasten rope around winch drum and lift winch assembly from tractor with rope and chain hoist.

d. Install Replacement Unit. Use reverse of removal procedure to install replacement winch assembly. Check winch sliding jaw clutch after winch is installed and clutch control rod is connected to

Figure 178—Removing Winch Assembly

make sure clutch jaws engage fully in jaws on drum when clutch lever is in engaged position, and that jaws have clearance to turn past each other when clutch lever is in disengaged position. To adjust, shorten or lengthen control rod by turning clevis at end of rod on or off.

142. REPLACEMENT OF WINCH DRIVE SHAFT ASSEMBLY.

a. Remove Drive Shaft Assembly. Tap shear pin out of front end of drive shaft and power take-off shaft (fig. 180). Loosen hexagon recessed set screw in universal connected to winch worm shaft (fig. 178). Remove two bolts that attach drive shaft bearing below rear of

TM 9-785
142

WINCH AND POWER TAKE-OFF

cab to side of hull frame (fig. 179). Tap universal at rear end of drive shaft off winch worm shaft as drive shaft and bearing assembly is lifted out of tractor.

b. **Install Replacement Assembly.** Lay assembly in hull of tractor as shown in figures 178 and 179. Hold center of assembly high enough to start rear universal onto winch worm shaft and front universal onto power take-off shaft. Line holes so shear pin can be driven

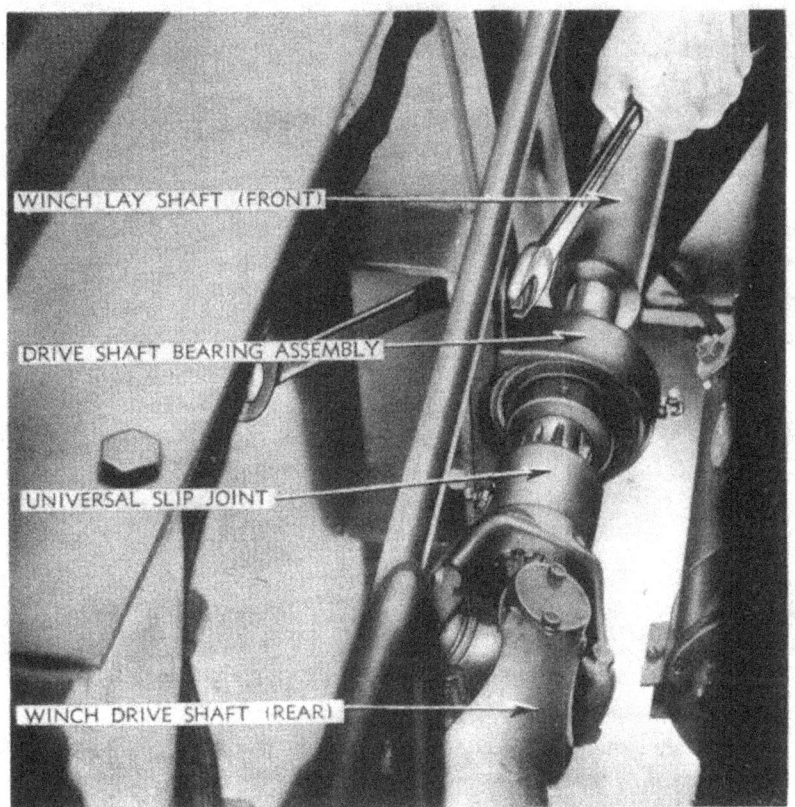

RA PD 307765

Figure 179—Removing Winch Drive Shaft Bearing Attaching Bolts

in to connect power take-off shaft and drive shaft universal. Tap shear pin into place (fig. 180). Tap rear universal onto winch worm shaft and tighten set screw in universal (fig. 178) against worm shaft. Attach drive shaft bearing bracket to bracket on side of hull frame (fig. 179) with two bolts with lock washers.

281

TM 9-785
18-TON HIGH SPEED TRACTOR M4

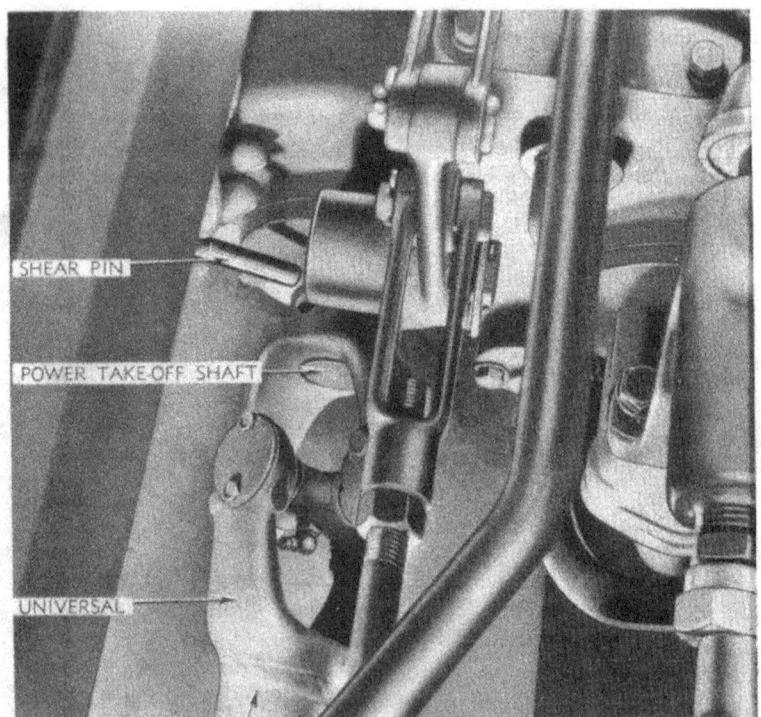

Figure 180—Removing Winch Shear Pin

143. REPLACEMENT OF POWER TAKE-OFF ASSEMBLY.

a. **Remove Power Take-off Assembly.**

(1) REMOVE TRANSMISSION OIL PUMP. Refer to paragraph 120 a for removal of pump.

(2) REMOVE POWER TAKE-OFF. Remove three cap screws holding left floor rail to side of hull. Remove two bolts holding line clips to left floor rail and remove spacers. Remove left floor rail. Disconnect transmission vent pipe at both ends. Remove clamp bolt from clip on filter in converter bleeder line (fig. 117) and remove transmission vent pipe. Remove pins to disconnect yoke on clutch control rods from lever at rear of power take-off housing (fig. 116). Unhook brake springs from bracket on front of power take-off (fig. 116). Remove pin to disconnect power take-off shifter rod from shifter shaft at front of power take-off housing. Remove pin to disconnect rear end of clutch control rod from lever on clutch control shaft in clutch housing. Re-

TM 9-785
143

WINCH AND POWER TAKE-OFF

move control rod. Remove shear pin from universal joint of winch lay shaft and power take-off shaft (fig. 180). Remove two bolts attaching winch lay shaft center bearing to side of hull at left side of engine (fig. 179). Drop shaft down and force universal joint from power take-off shaft. Remove pin connecting front end of clutch rod to clutch pedal and remove rod. Remove eight cap screws attaching power take-off to transmission case. Rotate top of power take-off assembly towards rear so that transmission oil pump attaching flange is toward bottom of hull. Lift power take-off up and out of place into rear seat compartment (fig. 181).

Figure 181—Removing Power Take-off Assembly

b. **Install Power Take-off Assembly.**

(1) INSTALL POWER TAKE-OFF. Cement gasket to attaching flange of power take-off housing. Lower assembly into place at side of transmission case (fig. 181) holding top of power take-off towards rear of tractor. Turn assembly right side up after it has been lowered into place, then mesh power take-off gear with transmission gear and attach assembly to transmission case with eight cap screws with lock washers. Install and connect front end of front clutch control rod to clutch pedal with yoke pin and cotter pin. Lift front end of winch drive shaft and tap universal onto power take-off shaft. Install shear pin (fig. 180). Attach winch drive shaft center bearing to bracket on side of hull frame with two bolts with lock washers (fig. 179). Install and

TM 9-785

18-TON HIGH SPEED TRACTOR M4

connect rear end of rear clutch control rod to lever on clutch control shaft in clutch housing with yoke pin and cotter pin. Connect shifter rod to shifter shaft at rear of power take-off with pin and cotter pin. Hook steering brake lever return springs into bracket on front of power take-off housing (fig. 116). Connect front end of rear clutch contol rod to lower hole in lever at rear of power take-off housing and rear end of front rod to upper hole in lever with yoke pins and cotter pins (fig. 116). Install transmission and differential vent pipe (fig. 117). Connect front end of pipe to fitting on power take-off, the rear end to connection at rear of cab. Install clip on vent pipe and filter in converter bleeder line as shown in figure 117. Install left floor rail to side of hull with three cap screws with lock washers. Attach two line clips to left floor rail with two bolts with lock washers.

(2) INSTALL TRANSMISSION OIL PUMP. Follow procedure in paragragh 120 b to install pump.

TM 9-785
144

Section XXIX

AMMUNITION AND CARGO BOXES

	Paragraph
Ammunition box	144
Cargo box	145

144. AMMUNITION BOX.

a. General. The ammunition box has ammunition storage racks to carry either 3-inch or 90-mm shells, and stowage compartments for other necessary equipment. Further description is given in paragraph 21. When this box is used on tractor, the fuel filler pipe "A" (fig. 185) is used on the tractor.

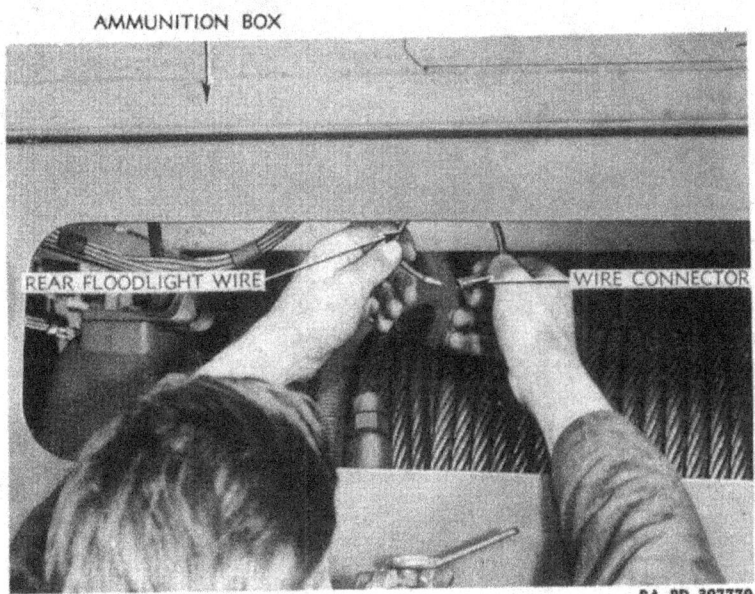

Figure 182—Disconnecting Rear Floodlight Wire

b. Replacement of Box.

(1) UNLOAD BOX AND DISCONNECT REAR FLOOD LIGHT WIRE. Remove canvas cover from top of box and remove contents from box. Disconnect rear flood light wire by pulling wire from connector inside hull under box above pintle (fig. 182).

(2) REMOVE BOLTS AND LIFT BOX FROM TRACTOR. Remove four bolts holding rear of box to hull. Open side doors of box and remove eight bolts holding front of box to hull and side fenders (fig. 183). Place a wood brace 4 x 4 x 92½ inches inside box, close to top against ends of box to prevent crushing in ends of box. Fasten a

18-TON HIGH SPEED TRACTOR M4

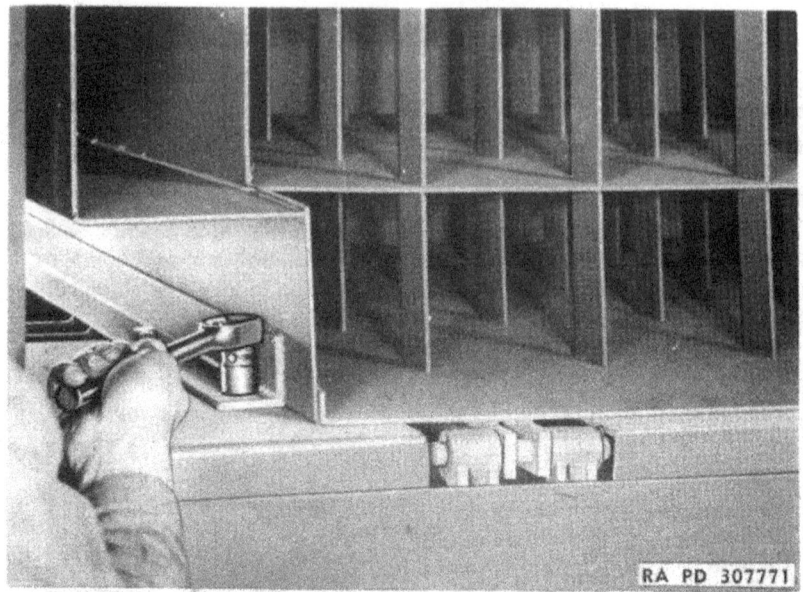

Figure 183—Removing Bolts Holding Ammunition Box

rope to box and hook rope into hook of chain hoist. Raise on hoist to raise rear of box about 2 inches so it will clear bed rail on rear of hull, then slide box back far enough to clear engine hood cross

Figure 184—Removing Ammunition Box from Tractor

AMMUNITION AND CARGO BOXES

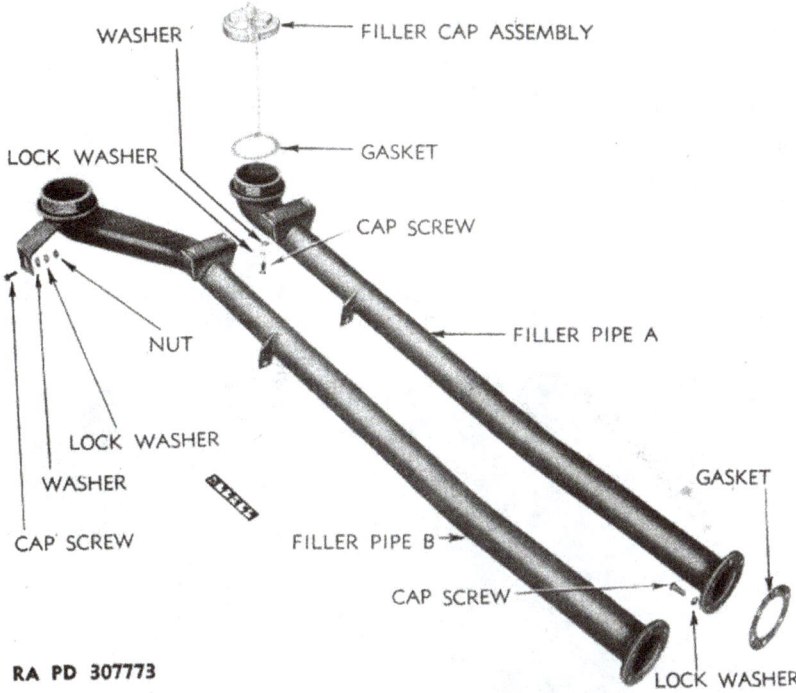

Figure 185—Fuel Tank Filler Pipes

frame at front of box. Raise box and remove it from tractor (fig. 184).

(3) INSTALL BOX. To install new box or reinstall the one removed, reverse the procedure for removal. Use lock washers on all bolts.

145. CARGO BOX.

a. **General.** The cargo box can be used in place of the ammunition box as both boxes are interchangeable. This box has facilities for carrying three different sizes of shells. Refer to paragraph 21 for complete description. When cargo box is installed on the tractor in place of the ammunition box, a fuel tank filler tube with an offset ("B" fig. 185) must be installed in place of the one used with the ammunition box.

b. **Replacement of Cargo Box.** Use same procedure as in paragraph 144 b to remove and install cargo box.

TM 9-785
146

18-TON HIGH SPEED TRACTOR M4

Section XXX

PINTLES

	Paragraph
Description	146
Replacement	147

146. DESCRIPTION.

a. The two different kinds of pintles furnished with the tractor are of the standard ordnance type and interchangeable on the tractor.

RA PD 307776

Figure 186—Removing 90-mm Pintle Assembly

They are exactly the same except that the one designed to pull the 90-mm and 3-inch antiaircraft guns has a universal swivel and yoke (fig. 186) while the one designed to pull the 155-mm, 8-inch Howitzer, and 240-mm guns has a hook for connection with the gun (fig.

288

AMMUNITION AND CARGO BOXES

11). A single large spring in the housing cushions shocks from starting, stopping, and towing of load.

147. REPLACEMENT.

a. Remove Pintle Assembly from Tractor. Removal procedure is the same for either 90-mm or 155-mm pintle. Remove the four bolts holding pintle assembly to rear of tractor (fig. 186) and remove it from tractor.

b. Install Pintle. Insert pintle into opening in rear of tractor. Plug for filling housing with lubricant should be towards right side of tractor. Secure pintle to tractor with four large bolts with lock washers (fig. 186). Tighten bolts securely.

TM 9-785
148-149

18-TON HIGH SPEED TRACTOR M4

Section XXXI

GUN RING, FIRE EXTINGUISHER, AND HULL DRAIN

	Paragraph
Gun ring	148
Fire extinguisher	149
Hull drain	150

148. GUN RING.

a. **Description.** The gun ring, located on top of cab over rear seat compartment, is of one piece welded construction. It provides for mounting of a gun skate for either the caliber .30 or caliber .50 machine gun which can be operated by a man standing in rear seat compartment. The gun can be turned in any direction or locked in

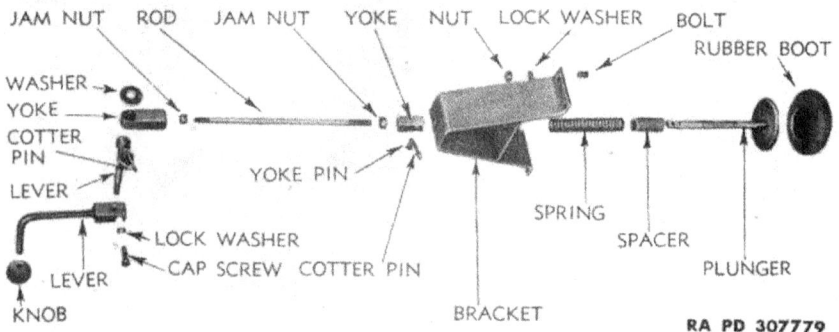

Figure 187—Hull Drain Assembly, Exploded

stationary position for traveling. A canvas cover is provided to cover gun ring.

b. **Replacement of Gun Ring.** Remove canvas cover from gun ring and remove gun skate if one is on ring. Remove the bolts attaching ring assembly to top of cab and lift ring assembly from tractor. Install replacement assembly in its place, using lock washers on all bolts.

149. FIRE EXTINGUISHER.

a. **Description.** Refer to paragraph 20 for description of extinguisher.

b. **Maintenance.** Determine contents by weight every six months. Recharge if weight of filled extinguisher is more than 6 ounces less than the weight stamped or written on body. Install new seal wire on trigger and write date of inspection or recharging on sticker.

GUN RING, FIRE EXTINGUISHER, AND HULL DRAIN

c. **Recharging.** Remove discharge horn and connect charging adapter to discharge pipe. Charge with four pounds of carbon dioxide (gas will open valve automatically). Release gas pressure in charging line and valve will close. Remove charging adapter, attach horn, and attach seal wire through holes in valve body and trigger. Enter weight on record.

150. HULL DRAIN.

a. **Description.** The hull drain is located in hull at right of transmission. It consists of a rubber valve which fits into a beveled opening in bottom of hull. The valve should remain closed except when it is desired to open it to drain water from hull. The operator opens valve with lever ahead of front seat (fig. 5).

b. **Replacement of Drain Assembly.** Remove pin from hull drain rod. Remove four bolts attaching bracket of drain valve assembly to floor of hull and lift entire assembly out of hull. Install new assembly in its place with four bolts with lock washers. Handle of lever should point toward front of tractor, when installed.

18-TON HIGH SPEED TRACTOR M4

PART THREE—STORAGE AND SHIPMENT

Section XXXII

SHIPMENT AND TEMPORARY STORAGE

	Paragraph
General instructions	151
Preparation for temporary storage	152
Loading and blocking for rail shipment	153

151. GENERAL INSTRUCTIONS.

a. Preparation for domestic shipment of the tractor is the same with the exception of minor added precautions as preparation for temporary storage. Preparation for shipment by rail includes instructions for loading the vehicle, blocking necessary to secure the vehicle on freight cars, number of vehicles per car, clearance, weight, and other information necessary to properly prepare the tractor for domestic rail shipment. For more detailed information and for preparation of the vehicle for indefinite storage refer to AR 850-18.

152. PREPARATION FOR TEMPORARY STORAGE.

a. Vehicles to be prepared for temporary storage are those ready for immediate service, but not used for less than 30 days. If vehicles are to be indefinitely stored after shipment by rail, they should be prepared for such storage at their destination.

b. If the vehicles are to be temporarily stored, take the following precautions.

(1) LUBRICATION. Lubricate the vehicle completely (par. 31).

(2) COOLING SYSTEM. If freezing temperature may normally be expected during the limited storage or shipment period, test the coolant with a hydrometer and add the proper quantity of antifreeze compound to afford protection from freezing at the lowest temperature anticipated during the storage or shipping period. Completely inspect the cooling system for leaks.

(3) BATTERY. Check battery and terminals for corrosion, and if necessary, clean and thoroughly service battery (par. 89).

(4) ROAD TEST. The preparation for limited storage includes a road test after the battery, cooling system, and lubrication service, to check the general condition of the vehicle. Correct any defects noted in the vehicle operation before the vehicle is stored, or note on a tag attached to the steering levers, stating the repairs needed or describing the condition present. Make a written report of these items to the officer in charge.

SHIPMENT AND TEMPORARY STORAGE

(5) FUEL IN TANKS. It is not necessary to remove fuel from the vehicle tanks for shipment within the United States, nor to label the tanks under Interstate Commerce Commission Regulations. Leave fuel in the tanks except when storing in locations where Fire Ordinances or other local regulations require removal of all gasoline before storage.

(6) EXTERIOR OF VEHICLE. Remove rust appearing on any part of the vehicle exterior with flint paper. Repaint painted surfaces whenever necessary to protect wood or metal. Coat exposed polished metal surfaces susceptible to rust, such as winch cables and chains, with medium grade preservative lubricating oil. Close firmly all doors, windows, windshields, and other openings. Top must be in place, raised and secured. Make sure paulins are in place and firmly secured. Leave rubber floor mats, when provided, in an unrolled position on the floor, not rolled or curled up. Equipment such as pioneer tools, track tools, and fire extinguishers can remain in place on the vehicle.

(7) INSPECTION. Make a systematic inspection just before shipment or temporary storage to insure all above steps have been covered and that the vehicle is ready for operation on call. Make a list of all missing or damaged items and attach it to the steering levers. Refer to Before-operation Service (par. 26).

(8) BRAKES. Release brakes and chock tracks.

c. **Inspections in Limited Storage.** Inspect tractors in limited storage for condition of battery, and in case of anticipated freezing weather, cooling system. If water is added to the battery when freezing weather is anticipated, recharge the battery with a portable charger or remove and charge the battery. Do not attempt to charge the battery by running the engine. If freezing temperature is expected, add the proper quantity of antifreeze compound to cooling system to afford protection from freezing.

153. LOADING AND BLOCKING FOR RAIL SHIPMENT.

a. **Preparation.** In addition to the preparation described in paragraph 152 when ordnance vehicles are prepared for domestic shipment, take the following steps.

(1) EXTERIOR. Cover the body of the vehicle with the canvas cover supplied as an accessory, or available for use during rail shipment.

(2) BATTERY. Disconnect the battery to prevent its discharge by vandalism or accident. This may be accomplished by disconnecting the positive lead, taping the end of the lead, and tying it back away from the battery.

(3) BRAKES. The brakes must be applied and the transmission placed in low gear after the vehicle has been placed in position with a brake wheel clearance of at least 6 inches ("A," fig. 188). Locate

TM 9-785
153
18-TON HIGH SPEED TRACTOR M4

the vehicles on the car in such a manner as to prevent the car from carrying an unbalanced load.

(4) MARKING CARS. All cars containing ordnance vehicles must be placarded "DO NOT HUMP."

(5) TYPES OF CARS. Ordnance vehicles may be shipped on flat cars, end door box cars, side door box cars, or drop end gondola cars, whichever type is most convenient.

b. Facilities for Loading. Whenever possible, load and unload vehicles from open cars under their own power, using permanent end ramps and spanning platforms. Movement from one flat car to another along the length of the train is made possible by cross-over plates or spanning platforms. If no permanent end ramp is available, an improvised ramp can be made from railroad ties. Vehicles may be loaded in gondola cars without drop ends by using a crane. In case of shipment in side door box cars, use a dolly type jack to warp the vehicles into position within the car.

c. Securing Vehicles. In securing or blocking a vehicle, three motions—lengthwise, sidewise, and bouncing—must be prevented. The following are approved methods of blocking and securing these vehicles on freight cars.

(1) METHOD ONE. Place four blocks ("B," fig. 188), one to the front and one to the rear of each track. Nail the heel of each block to the car floor with five 40-penny nails. Toenail to the car floor with two 40-penny nails that portion of each block which is under the track. Locate two blocks ("C") on each side of the vehicle on the outside of each track. Nail each block to the car floor with three 40-penny nails. These blocks may be located on the inside of the tracks if conditions warrant. Pass four strands, two wrappings, of No. 8 gage black annealed wire over the axle of each inside bogie wheel and secure to the nearest stake pocket (fig. 188). Tighten the wire enough to remove slack. This strapping is not required when gondola cars are used, but when a box car is used, this strapping must be applied in a similar fashion and attached to the floor by use of blocking or anchor plates.

(2) METHOD TWO. Place two blocks ("F," fig. 188), one to the front and one to the rear of the tracks. These blocks are to be at least as long as the over-all width of the vehicle at the car floor. Locate eight blocks ("G") against the blocks ("F") to the front and to the rear of each track. Nail the lower block to the floor with three 40-penny nails and the top block to the lower block with three 40-penny nails. Locate and secure blocks ("C") and wire strapping as explained in Method One.

SHIPMENT AND TEMPORARY STORAGE

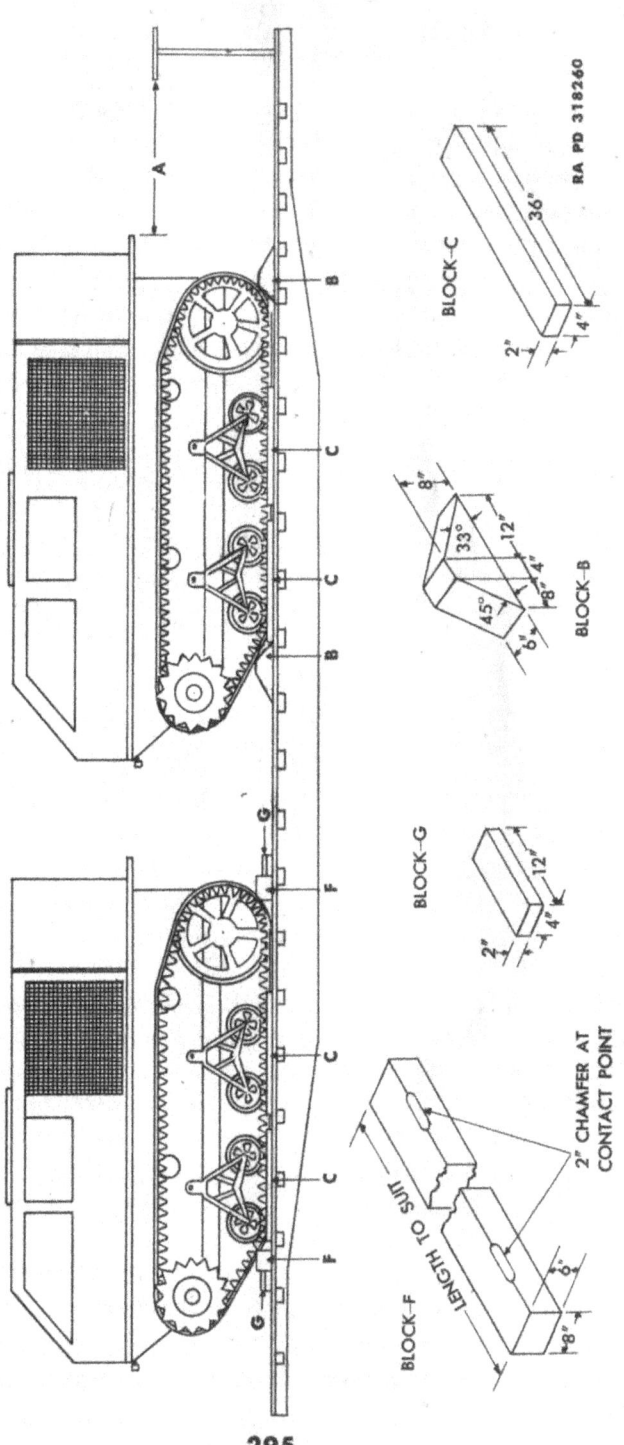

Figure 188—Blocking Requirements for Rail Shipment

18-TON HIGH SPEED TRACTOR M4

Length (over-all) ....16.92 ft
Width ....8.17 ft
Height ....8.25 ft
Area of car floor occupied per vehicle......138 sq ft
Volume occupied per vehicle.............1,140 cu ft
Shipping weight per vehicle...............31,400 lb
Bearing pressure (lb per sq ft).............229 lb

TM 9-785

REFERENCES

STANDARD NOMENCLATURE LISTS.

Tractor, high speed, 18-ton	SNL G-150
Cleaning, preserving and lubrication materials, recoil fluids, special oils, and miscellaneous related items	SNL K-1
Soldering, brazing, and welding materials, gases, and related items	SNL K-2
Tool Sets—motor transport	SNL N-19
Interchangeability chart of organizational special tools for combat vehicles	SNL G-19
Current Standard Nomenclature Lists are listed above. An up-to-date list of SNL's is maintained in the Index to Ordnance Publications	OFSB 1-1

EXPLANATORY PUBLICATIONS.

Military motor vehicles	AR 850-15
List of publications for training	FM 21-6

Automotive Materiel.

Automotive electricity	TM 10-580
Electrical fundamentals	TM 1-455
Fuels and carburetion	TM 10-550
The internal combustion engine	TM 10-570
The motor vehicle	TM 10-510
Tune-up and adjustments	TM 10-530

Care and Preservation.

Automotive lubrication	TM 10-540
Cleaning, preserving, lubricating, and welding materials and similar items issued by the Ordnance Department	TM 9-850
Explosives and demolitions	FM 5-25
Motor vehicle inspections and preventive maintenance services	TM 9-2810

Decontamination.

Chemical decontamination materials and equipment	TM 3-220
Decontamination of Armored Force vehicles	FM 17-59
Defense against chemical attack	FM 21-40

TM 9-785

18-TON HIGH SPEED TRACTOR M4

Storage and Shipment.

Registration of motor vehicles............... AR 850-10

Rules governing the loading of mechanized and motorized army equipment, also, major caliber guns, for the United States Army and Navy, on open top equipment published by Operations and Maintenance Department of Association of American Railroads.

Storage of motor vehicle equipment........... AR 850-18

Ordnance field service storage and shipment Chart—group G major items............... OSSC-G

TM 9-785

INDEX

Page No.

A

Air pressure gage 193
Air trailer brake controls
  air brake application valves
    hand-controlled air brake
      valve replacement 244
    pedal-controlled air brake
      valve replacement 244
  air compressor
    description 237
    maintenance 238
    replacement 242
  air pressure governor
    adjustment
      air compression 242
      procedure 243
  description
    components of air brake
      control system 235
    operation 236
    purpose 235
  tests for air brake system
    air pressure tests 245
    frequency 245
    leakage tests 246
    operation tests 246
    valve delivery pressure tests. 246
Ammeter 195
Ammunition and cargo boxes
  general 285, 287
  replacement 285, 287
Ammunition boxes and shell hoist 26

B

Battery 165
Brake operation
  air brakes 18
  electric brakes 18
Brakes, steering 224

C

Carburetors 133
Clutch, master 205
Cold-weather operation 30

D

Data, tabulated
  distributor 127
  engine 110
  generator regulator 172

Page No.

spark plugs 127
vehicle 9
Description
  air cleaner (oil bath)......... 142
  air compressor 237
  air precleaner 140
  air pressure gage 193
  air trailer brake controls...... 235
  ammeter 195
  battery 165
  bogie assemblies 267
  carburetors 133
  cooling fan and fan drive
    assembly 157
  cranking motor button switch.. 176
  cranking motor switch
    (magnetic) 175
  distributor 127
  electric trailer brake controls.. 247
  engine 108
  engine cooling system 145
  engine lubricating oil cooler... 163
  engine lubricating oil filters... 162
  engine lubricating oil pressure
    gage 196
  engine shut-off 136
  engine tachometer 198
  engine temperature gage 198
  exhaust manifold 115
  fire extinguisher 23
  fuel filter 131
  fuel gage 198
  fuel pump 132
  generator 168
  generator regulator 170
  governor 112
  gun ring 290
  hour meter 19, 200
  hull drain 291
  ignition system 127
  intake and exhaust systems.... 139
  lights 178
  low air pressure indicator...... 201
  master clutch 205
  muffler 144
  primer pump 135
  propeller shaft 208
  radiator assembly 150
  siren 187
  speedometer 202

299

18-TON HIGH SPEED TRACTOR M4

D—Cont'd

Description—Cont'd
- starter (cranking motor) 174
- steering brakes 224
- thermostat 149
- torque converter 209
- torque converter fluid filter.... 218
- torque converter pressure gage. 203
- torque converter fluid
  - temperature gage 204
- track assemblies 255
- track drive sprockets 261
- track support roller assemblies. 273
- trailing idler assemblies 263
- transmission, differential, and
  - final drives 220
- transmission oil pressure gage.. 204
- transmission oil temperature
  - gage 204
- vehicle 7
- water pump 146
- winch 275
- windshield wipers 187
- wiring system 187

Description of vehicle
- engine 7
- equipment 7
- general 7
- seats 7
- steering 7
- tabulated data
  - capacities 10
  - dimensions 9
  - engine 9
  - general 9
  - performance 9
  - power take-off 10
  - steering 10
  - tracks 10
  - winch 10

Driving controls and operation
- engine starting instructions.... 12
- hour meter
  - description 19
  - how to read meter 19
- inspection of new tractor 11
- instruments
  - air pressure gage 21
  - low air pressure indicator... 21
  - ammeter 21
  - engine oil pressure gage..... 21
  - fuel gage 21
  - general 21
  - speedometer 21
  - tachometer 21
  - torque converter fluid
    - pressure gage 22
  - torque converter fluid
    - temperature gage 22
  - transmission oil pressure gage 21
  - transmission oil temperature
    - gage 21
- lighting system 19
- operation of new tractor 12
- parking vehicle 19
- preparing new tractor for use.. 11
- steering levers and brakes..... 16
- stopping engine 14
- towing vehicle
  - to start engine 17
  - towing disabled vehicle 17
- use of master clutch and
  - gearshift lever 14
- use of trailer brakes
  - connections 18
  - operation of air brakes 18
  - operation of electric brakes.. 18
  - types of trailer brake controls 18

E

Electrical system
- battery
  - adding water 166
  - description 165
  - hydrometer readings 165
  - idle batteries 167
  - replacement of battery 168
  - specific gravity 165
  - table of electrolyte
    - temperatures 166
  - terminals 167
  - vent plugs 167
  - general 165
- generator
  - description and data 168
  - maintenance 169
  - operation 169
  - replacement 169
- generator regulator
  - description and data 170
  - maintenance 172
  - replacement of regular
    - assembly 174

INDEX

E—Cont'd

Electrical system—Cont'd
light switches
  general 181
  main light switch replacement 183
  panel light or rear floodlight
    switch replacement 184
  stop light switch replacement 187
lights
  blackout driving light
    replacement 180
  description of lighting system 178
  light bulb replacement 180
  replacement of headlight or
    rear floodlight 178
  service stop and taillight
    replacement 180
siren
  description 187
  replacement 187
starter (cranking motor)
  cranking motor button switch 176
  cranking motor switch 175
  description and data 174
  maintenance 174
  replacement of cranking
    motor 175
windshield wipers
  description 187
  replacement 187
wiring system
  description 187
  repair and replacement of
    wires 192
Electric trailer brake controls
  control unit replacement
    brake controller assembly.... 253
    coupling socket 254
    load control unit 253
    resistor unit 253
    wiring system 254
  description
    brake controller 249
    coupling socket 249
    general 247
    load control 247
    resistor 247
  operation 252

Engine
  accessories 109
  cylinder head gaskets 115
  data, tabulated 110
  general description 108
  governor 113
  installation 123
  intake and exhaust manifolds.. 113
  lubrication 109
  removal of engine 116
  valve adjustment 111
Engine cooling system
  cooling and fan drive assembly
    description 157
    replacement 158
  description 145
  draining system 146
  filling system 145
  radiator assembly
    assembly 157
    description 150
    disassembly 155
    flushing radiator 151
    installation 157
  thermostat
    description 149
    replacement 149
  water pump
    description 146
    installation 148
    removal 146
Engine lubrication oil filters and oil
  cooler
  engine lubricating oil cooler
    description 163
    replacement of oil cooler
      assembly 163
  engine lubricating oil filters
    description 162
    replacement of elements 162
    replacement of filter assembly 162
    general 160
Engine tachometer 198
Engine temperature gage 198

F

Fire extinguisher 23
First echelon preventive maintenance
  after-operation and weekly
    service 42

TM 9-785

18-TON HIGH SPEED TRACTOR M4

F—Cont'd

First echelon preventive maintenance—Cont'd
- at-halt service 41
- before-operation service 37
- during-operation service 39
- purpose 36

Fuel system
- carburetor control adjustment
  - choke 134
  - idling screws 134
  - throttle 134
- carburetors
  - adjustment 134
  - description 133
  - replacement 133
- engine shut-off
  - air valve adjustment 136
  - air valve assembly
    - replacement 137
    - description 136
  - ignition switch replacement.. 138
- fuel filter
  - description 132
  - service and replacement 131
- fuel pump
  - description 132
  - replacement 133
- fuel tank
  - general 131
  - sediment sump 131
- general 131
- primer pump
  - description 135
  - replacement 136

G

Gun equipment and gun tools.... 57
Gun ring, fire extinguisher, and hull drain
- fire extinguisher
  - description 290
  - maintenance 290
  - recharging 291
- gun ring
  - description 290
  - replacement 290
- hull drain
  - description 291
  - replacement of assembly 291

H

Hour meter, how to read 19

I

Ignition system
- description 127
- distributor
  - description and data 127
- ignition timing 129
- spark plugs 127

Installation procedure
- air compressor 148, 242
- air compressor belt idler assembly 148
- air pressure gage 193
- ammunition box 126, 287
- bogie components 272
- brake controller assembly 254
- brake housing covers 279
- carburetor 133
- carburetor assembly 114
- clutch assembly 207
- cylinder head gasket 116
- differential cover and gear shifter lever guide 233
- distributor 129
- engine 123
- engine hood and frame 125
- exhaust manifold 115
- generator 169
- hour meter 201
- idler assembly 266
- manifold 114
- muffler 144
- oil and water lines 148
- pintle 289
- power take-off assembly 283
- pump 148
- radiator and supporting frame.. 124
- radiator assembly 157
- radiator expansion tank 124
- regulator assembly 174
- rocker arm covers and distributor cap 112
- sprockets 263
- steering brakes 231
- thermostat assembly 150
- torque converter 217
- track assembly 258
- transmission oil pump.....222, 284

302

INDEX

I—Cont'd

	Page No.
Installation procedure—Cont'd	
valve assembly	244
winch assembly	281
Instruments and gages	
air pressure gage	
description	193
replacement	193
ammeter	
description	195
replacement	195
engine lubricating oil pressure gage	
description	196
replacement of gage	196
engine tachometer	
description	198
replacement of tachometer	198
replacement of tachometer drive shaft	198
engine temperature gage	
description	198
replacement of gage	198
replacement of gage operating unit	198
fuel gage	
description	198
replacement of gage	198
replacement of operating unit	200
general	193
hour meter	
description	200
replacement	200
low air pressure indicator	
description	201
replacement of bulb	201
replacement of low air pressure indicator switch	201
speedometer	
description	202
replacement	202
replacement of speedometer drive shaft	203
torque converter fluid pressure gage	
description	203
replacement of gage	203
replacement of gage operating unit	203

	Page No.
torque converter fluid temperature gage	
description	204
replacement of gage	204
replacement of thermal unit	204
transmission oil pressure gage	
description	204
replacement of gage	204
replacement of gage operating unit	204
transmission oil temperature gage	
description	204
replacement of gage	204
replacement of thermal unit	204
Intake and exhaust systems	
air cleaner (oil bath)	
description	142
replacement	144
service	143
air precleaner	
description	140
replacement	141
service	140
general description	
air intake system	139
exhaust system	139
muffler	
description	144
installation	144
removal of assembly	144
Instruments on vehicle	21
Introduction	
arrangement of manual	3
scope of manual	3

L

Lighting system	19
Lights	178
Loading and blocking for rail shipment	
facilities for loading	294
preparation	
battery	293
brakes	293
exterior	293
marking cars	294
types of cars	294
securing vehicles	294
Low air pressure indicator	201
Lubrication	
introduction	47

TM 9-785

18-TON HIGH SPEED TRACTOR M4

L—Cont'd   Page No.   Page No.

Lubrication—Cont'd
  lubrication guide
    air cleaners 47
    bogie wheels and track support
      rollers 50
    crankcase 47
    cranking motor 50
    engine oil filters 50
    fittings 47
    fuel filter 50
    gear cases 50
    general 47
    oilcan points 51
    points requiring no lubrication
      service 51
    supplies 47
    torque converter fluid filters.. 50
    torque converter fluid
      reservoir 50
    trailing idlers 50
    universal joints and slip joint 50
  points to be serviced and/or
    lubricated by ordnance
    maintenance personnel
      generator 51
    tachometer and speedometer
      flexible drive shafts 51
  reports and records 51
Lubrication guide 47

M

Maintenance allocation
  allocation of maintenance 59
  echelon system of allocation
    air brake group 65
    clutch assembly 60
    cooling group 61
    electric brake group 66
    electrical group 61
    engine (Waukesha, 145GZ). 62
    exhaust group 63
    final drive assemblies 61
    fire extinguisher 64
    frame and body group 64
    gooseneck 66
    instruments 64
    power take-off case 66
    propeller shafts 64
    suspension group 64
    torque converter assembly... 61

    transmission and differential
      assembly 65
  scope 59
Manual
  arrangement of 3
  scope of 3
Master clutch
  clutch pedal adjustment
    adjustment of free pedal
      travel 205
    general 205
  description 205
  replacement of clutch drive disk 206

O

Operation of auxiliary equipment
  ammunition boxes and shell hoist 26
  fire extinguisher
    description 23
    operation 23
  tire inflation hose 24
  winch and controls
    to attach cable to load...... 23
    to pull load 23
Operation under unusual conditions
  desert operation 34
    electrical difficulties 35
    excessive wear due to sand.. 35
    overheating 35
  extreme cold weather
    antifreeze in cooling system.. 30
    batteries 30
    cold-weather starting 31
    condensation 30
    how to connect heaters 34
    lubricants 30
    stopping tractor 30
    use of heaters and covers 31
  water or mud 34
Organization tools and equipment 83

P

Pintles
  description 288
  installation 289
  removal 289
Preventive maintenance
  first echelon services
    after-operation and weekly
      service 42
    at-halt service 41
    before-operation service 37

INDEX

P—Cont'd

	Page No.
Preventive maintenance—Cont'd	
first echelon services—cont'd	
during-operation service	39
purpose	36
second echelon services	67
road test chart	69
Propeller shaft	208

R

	Page No.
References	
explanatory publications	
automotive materiel	297
care and preservation	297
decontamination	297
storage and shipment	298
standard nomenclature lists	297
Removal procedure	
air compressor	147
air compressor belt idler assembly	147
air pressure gage	193
ammunition or cargo box	116, 285
bogie components	268
brake controller assembly	253
brake linkage	230
brake shoes	230
carburetor assembly	113
clutch assembly	206
contact points	129
cover plate, vent pipe, and fan shroud	155
cylinder head and gasket	116
differential cover	227
distributor	128
drive shaft assembly	280
engine	122
engine hood and frame	116
engine lubricating oil cooler element	163
engine lubricating oil cooler housing	163
exhaust manifold	115
fan blade assembly	158
fan drive gear assembly	158
fan shaft and housing assembly	158
generator	169
hour meter	200
idler assembly	264
idler yoke and support bracket assembly	264
intake manifold	113
intake manifold assembly	115
lower inspection plate	147
manifold	113
muffler assembly	144
pintle assembly	289
power take-off assembly	282
propeller shaft assembly	213
radiator and supporting frame	118
radiator assembly	154
radiator expansion tank	118
radiator mounting cap screws	153
regulator assembly	174
seat back cushion	153
thermostat assembly	150
tool box	163
torque converter	215
torque converter fluid filter	215
track assembly	255
track drive sprockets	261
transmission oil pump	222
transmission oil radiator	156
volute spring assembly	270
water and torque converter fluid radiators	157
water outlet manifold	116
water pump	146
winch assembly	280
wires from main light switch	184

S

	Page No.
Second echelon preventive maintenance	
first echelon participation	67
frequency	67
general procedures	67
road test chart	69
maintenance operations	72
specific procedures	69
Shipment and temporary storage	
general instructions	292
inspections in limited storage	293
preparation for temporary storage	
battery	292
cooling system	292
exterior of vehicle	293
fuel in tanks	293
inspection	293
lubrication	292
road test	292
Speedometer	202
Steering brakes	
brake adjustment	224
description	224

TM 9-785

TM 9-785

18-TON HIGH SPEED TRACTOR M4

	Page No.		Page No.

Steering brakes—Cont'd
  installation 231
  steering brake replacement.... 227

T

Tabulated data
  distributor 127
  engine 110
  generator regulator 172
  spark plugs 127
  vehicle 9
Thermostat 149
Tire inflation hose 24
Tools and equipment stowage on tractor
  care of equipment 58
  equipment 55
  general 52
  gun equipment 57
  gun spare parts 57
  gun tools 57
  spare parts 55
  vehicle tools
    pioneer 52
    vehicular 52
Torque converter
  auxiliary fluid pump 211
  cooling system 211
  fluid cooling radiator 218
  fluid filter 218
  freewheel assembly 212
  maintenance 212
  principles of operation 210
  purpose and location 209
  replacement 213
  reserve tank 211
Torque converter and propeller shaft
  propeller shaft
    description 208
    propeller shaft assembly replacement 208
  torque converter
    description 209
    maintenance 212
    replacement 213
    replacement of auxiliary fluid pump 217
  torque converter fluid cooling radiator
    general 218
    replacement 219

torque converter fluid filter
  description 218
  filter element replacement... 218
  maintenance 218
Torque converter fluid pressure gage 203
Towing vehicle
  to start engine 17
  towing disabled vehicle 17
Tracks and suspensions
  bogie assemblies
    bogie component installation. 272
    bogie component removal... 268
    bogie wheel replacement.... 268
    description 267
    maintenance 268
  general 255
  idler assembly installation 266
  idler yoke removal 264
  track assemblies
    adjustment 260
    description 255
    installation 258
    maintenance 255
    removal 255
  track drive sprockets
    description 261
    installation 263
    removal 261
  track installation and adjustment 267
  track support roller assemblies
    description 273
    maintenance 273
    replacement 274
  trailing idler assemblies
    description 263
    maintenance 264
    removal 264
Transmission, differential, and final drives
  description 220
  lubrication 220
  transmission oil cooling radiator replacement 223
  transmission oil pump replacement 220
Trouble shooting
  cooling system 92
  electrical system 107

INDEX

T—Cont'd

	Page No.
Trouble shooting—Cont'd	
engine	
instructions	84
tests to determine mechanical condition	87
trouble shooting chart	85
fuel system	91
generator and regulator	
excessive output	94
no generator output	94
noisy generator	94
tests	94
unsteady or low output	94
ignition system	89
introduction	84
lubrication system	
detection of oil leaks	93
excessive oil consumption	93
sudden loss of oil pressure	93
power train	95
excessively high transmission oil temperature	100
inspection	99
lack of transmission oil pressure	99
oil leaks	100
steering brakes	99
torque converter	100
starting system	88
tracks and suspensions	102
trailer brake controls	104
air brake	104
air brake system test	106
electric brake control system	106
winch and power take-off	103

V

	Page No.
Valve adjustment	111
Vehicle tools	52

W

	Page No.
Winch and controls	23
Winch and power take-off	
description	
power take-off	275
winch	275
maintenance	275
power take-off assembly replacement	
installation	283
removal	282
winch adjustments	
automatic brake	275
winch jaw clutch	279
winch assembly replacement	279
winch drive shaft assembly replacement	280
Wiring system	187

[A. G. 300.7 (26 Mar. 43)]

BY ORDER OF THE SECRETARY OF WAR:

G. C. MARSHALL,
Chief of Staff.

OFFICIAL:

J. A. ULIO,
Major General,
The Adjutant General.

DISTRIBUTION: R9 (4); Bn9 (2); C9 (8); IC and H 4, 6, and 44 (5).

(For explanation of symbols, see FM 21-6.)

Also Now Available!

Visit us at:

www.PeriscopeFilm.com

IN HIGH DEFINITION
NOW AVAILABLE!

COMPLETE LINE OF WWII AIRCRAFT FLIGHT MANUALS

WWW.PERISCOPEFILM.COM

©2011 Periscope Film LLC
All Rights Reserved
ISBN #978-1-937684-96-9

www.ingramcontent.com/pod-product-compliance
Lightning Source LLC
Chambersburg PA
CBHW060339170426
43202CB00014B/2815